Electrodynamics of Superconductors

Superconductivity is a remarkable, on the verge of miraculous, macroscopic quantum phenomenon with a boundless potential for a wide variety of applications. Over the last century the physics of superconductivity constitutes a major part of research in condensed matter and yet the electrodynamics of superconductors appears as only a minor part in textbooks on electrodynamics and superconductivity.

The book offers a fresh look at electrodynamics of continuous media with detailed description of thermal and electromagnetic properties of superconductors, emphasising physical meaning of concepts and principles without excessive mathematics. To facilitate understanding, it is accompanied by problems and worked solutions.

The book will be of interest to senior undergraduate and graduate students of physics and chemistry who have taken a calculus-based course in general physics. It is a valuable reference for researchers involved in studies of superconductivity, the physics of metals, and condensed matter physics.

Key Features:

- The first book - in terms of semi-classical physics - addressing both equilibrium and nonequilibrium, thermal and transport properties of superconducting materials.
- Presents an accessible overview without excessive mathematics.
- Accompanied by problems with solutions to aid understanding.

Vladimir Kozhevnikov is a retired professor of Tulsa Community College, USA; over the last two decades he was also a vising researcher at the KU Leuven, Belgium. He received his PhD from the Moscow Aviation Institute (MAI) and D. Sc. (habilitation) in physics and mathematics from the Kurchatov Institute of Atomic Energy (KIAE), Russia. Before moving to the US in 1996 he was physics professor at MAI and a senior researcher at the KIAE. His researches at that time were mostly focused at the liquid-gas criticality, metal-dielectric transition and wetting in fluid metals. He was the recipient of the 'Kurchatov Prize' in 1996. His current research interests primarily include superconductivity, phase transitions, and thermodynamics in magnetic fields. Dr. Kozhevnikov has published a book on thermodynamics of magnetized materials and superconductors and over 70 papers in referred journals.

Electrodynamics of Superconductors

Vladimir Kozhevnikov

CRC Press
Taylor & Francis Group
Boca Raton London New York

CRC Press is an imprint of the
Taylor & Francis Group, an **informa** business

Designed cover image: Vladimir Kozhevnikov

First edition published 2025
by CRC Press
2385 NW Executive Center Drive, Suite 320, Boca Raton FL 33431

and by CRC Press
4 Park Square, Milton Park, Abingdon, Oxon, OX14 4RN

CRC Press is an imprint of Taylor & Francis Group, LLC

ISBN: 978-1-032-40644-2 (hbk)
ISBN: 978-1-032-41008-1 (pbk)
ISBN: 978-1-003-35578-6 (ebk)

DOI: 10.1201/9781003355786

Typeset in CMR10 font
by KnowledgeWorks Global Ltd.

Publisher's note: This book has been prepared from camera-ready copy provided by the authors.

To Polina and Asya

Contents

Preface

> We should be grateful to God that
> He created the world such that all simple
> is true, and all complicated is untrue.
>
> H. Skovoroda[1]
>
> From curves I acknowledge only straights.
>
> D. Arnold[2]

Over its century-long history, superconductivity has undergone three major shocks. The first shock was the very discovery of this seemingly impossible phenomenon by Kamerlingh Onnes. Suffice it to say that at the first presentation of superconductivity at the Solvay Conference in 1911 (nine out of the 23 participants were Nobel laureates) Paul Langevin was the only one who asked the question.

The second shock was the discovery of the Meissner effect by Meissner and Ochsenfeld, and Rjabinin and Shubnikov in 1933−34, when it turned out that superconductors are also super-diamagnets. These shocks were so strong that Ashcroft and Mermin in their classical textbook on solid state physics (1976) noted that a chapter on superconductivity is "largely self-contained".

And the last (for today) shock was the discovery of high-temperature superconductors by Müller and Bednorz in 1986, when it turned out that none of the existing theories is appropriate to describe their properties. This shock is still ongoing with no apparent prospect of exiting the tunnel.

There have been many similar situations in the history of science, and each time the solution found proved to be very simple. One can recall, for example, thousand-year observations of celestial stars and mysterious loops in the trajectories of planets vs the Newton law, or the history of research on magnetism and electricity vs the Faraday law, and lots more. The case of superconductivity appeared much easier: it does not require the introduction of new laws. As the reader will see, the properties of all superconductors stem from the laws of classical physics supplemented by the Bohr-Sommerfeld quantization condition for paired electrons with mutually compensated spins (Cooper pairs). This fact allows the phenomenon of superconductivity to be described in an

---

[1] Hryhoriy Skovoroda (1722−1794) was a Ukrainian philosopher. Quoted from P. L. Kapitza, *Experiment, Theory, Practice* (M., Nauka, 1974).

[2] Dmitry I. Arnold (1939−2019) was a Soviet/Russian physicist.

easily accessible mathematical language with a clear-cut physical meaning. This is the language employed in this book.

As a rule, capital characters are used for quantities related to an entire body under consideration and lowercase symbols for the values related to its unit volume. Unfortunately, I was unable to avoid using the same symbol for different quantities. However, since such quantities are not met simultaneously, this should not lead to confusion.

Units of the *cgs* system are used, because these well-balanced units are the most appropriate for discussing electrodynamics of continuous media. The different dimensions of induction $\mathbf{B}$ and intensity $\mathbf{H}$ of the magnetic field in the SI, make this system inapplicable for this purpose. No duplication of formulas in SI units is offered, since this complicates reading and distorts physical meaning. The latter will become particularly clear when discussing vector potentials.

The book is intended for research students specializing in relevant areas of physics and chemistry. It can also be recommended to senior undergraduates, well acquainted with the calculus-based course of general physics. The book will be of interest to researchers involved in studies of superconductivity, physics of metals, and condensed matter physics and chemistry in general. It is also recommended to instructors of physics and chemistry in high schools and colleges, as well as to everyone interested in magnetism, condensed matter and, of course, superconductivity.

I am deeply beholden to the late Prof. Vladimir Kresin (1934−2022) for extremely valuable criticism and encouragement at an early stage of this project. I am obliged to Prof. Orest Symko for comments that greatly stimulated this work. My deepest thanks to Prof. Joseph Indeceu and Prof. Chris Haesendonck for hosting my visits to the University of Leuven, without which this book would not have had a chance. A sea of gratitude to Prof. Valery Cherepanov, Prof. Oscar Bernal and Dr. Mariela Menghini for reading and commenting on the manuscript. I am grateful to Prof. Rinke Wijngaarden and Dr. Mariela Menghini for sharing their wonderful video recordings. Last but not least, I am indebted to my wife Yelena Spivak for her patience and support throughout all stages of this project.

Vladimir Kozhevnikov,
July 2024.

Frequently used symbols and abbreviations

Symbols

$\mathbf{A}$ vector potential of the magnetic field intensity

$\mathbf{A}_B$ vector potential of the magnetic field induction $\mathbf{B}$

$\mathbf{B}$ induction of magnetic field (magnetic induction)

B_n, B_t normal and tangential components of the induction

C heat capacity

c speed of light and specific heat capacity

D, D_n, D_s period and widths of the normal and superconducting domains of the laminar structure of the IS, respectively

$\mathbf{E}$ electric field

E_c condensation energy

$\mathbf{E}_i$ induced electric field

E_k kinetic energy

E_m magnetic energy

$\mathcal{E}$ thermo-e.m.f. (Seebeck voltage)

e electron charge

$\mathbf{e}$ microscopic electric field

e_c condensation energy density

e_{cp} bonding energy of Cooper pairs

$F(T, V, \mathbf{B})$, $\widehat{F}(T, V, \mathbf{H})$, $\widetilde{F}(T, V, \mathbf{H}_0)$ different forms of free energy

$\mathbf{F}_L$, $\mathbf{F}_E$, $\mathbf{F}_M$ Lorentz, electric and magnetic forces, respectively

g g-factor

$\mathbf{g}_b$, $\mathbf{g}_t$ linear density of a bound and total surface current

$\mathbf{H}$ magnetic field intensity (strength)

$\mathbf{H}_B$ the Barnett field

H_n, H_t normal and tangential components of the intensity

$\mathbf{H}_0$ applied field intensity

H_c thermodynamic critical field

H_{c1}, H_{c2} lower and upper critical field, respectively

H_{c3} critical field of superconductivity nucleation

H_{cr} critical field of the S/N transition

H_d demagnetizing field intensity

$\mathbf{h}$ microscopic magnetic field

h Planck constant

$\mathbf{I}$ magnetization

J electric current

J_c critical current

J_{cp} current in a Cooper pair induced by a magnetic field

J_Q heat flow (current)

J_S entropy current

$\mathbf{j}$ time-mean density of microscopic current

$\mathbf{j}_{mic}$, $\mathbf{j}_{mol}$, $\mathbf{j}_c$ bulk density of a microscopic, molecular and conduction currents, respectively

k_B Boltzmann constant

L_h healing length

$\mathbf{L}$ angular momentum

l_m mean free path

$\mathbf{M}$, $\mathbf{M}_a$ specimen and atomic magnetic moment, respectively

m electron mass

m_{cp} Cooper pair mass

$\mathbf{n}$ unit vector normal to specimen surface

n_s, n_{cp} number density of superconducting electrons and Cooper pairs, respectively

$\mathbf{p}$ linear momentum

R_0 radius of an electron orbit in Cooper pairs

r_i root mean square (rms) radius of bound currents induced in atoms and Cooper pairs

Q latent heat

S entropy

S_s, S_n entropy of conduction electrons in the S and N states

S_m magnetic part of specimen entropy

$\mathbf{s}$ spin angular momentum

T temperature and kinetic energy

T_c critical temperature

t time

U internal energy

V volume

$\mathbf{v}_{mic}$ linear velocity of microscopic motion

$\mathbf{v}_0$ linear velocity of microscopic motion in zero field

$\mathbf{v}_i$ linear velocity of induced motion

W work

א parameter aleph

α S/N surface tension; the fine-structure constant

δ domain-wall energy parameter

ϵ_{cp} kinetic energy of Cooper pairs

$\Delta\Phi$ total phase difference of induced motion of Cooper pairs in weakly coupled
      superconductors

γ gyromagnetic ratio

η demagnetizing factor

ι microscopic angular momentum (iota)

κ Ginzburg-Landau parameter

λ penetration depth

λ_{eff} effective penetration depth

λ_L London penetration depth

$\boldsymbol{\mu}$ microscopic magnetic moment

$\boldsymbol{\mu}_{cp}$ induced magnetic moment per Cooper pair

μ_m magnetic permeability

ξ coherence length

o (omicron) Larmor angular frequency (velocity)

σ surface density of electric charges

τ gyroscopic torque

Φ magnetic flux

Φ_0 superconducting magnetic flux quantum

χ magnetic susceptibility per unit volume

Ψ Ginzburg-Landau order parameter

Ψ_M magnetic scalar potential of the **H**-field

φ phase of induced motion of Cooper pairs

$\boldsymbol{\Omega}$ angular velocity

$\boldsymbol{\omega}_r$ angular velocity of rotation

$\boldsymbol{\omega}_i$ angular velocity of induced circular motion

Abbreviations

AMMS   averaged model of the mixed state

GL   Ginzburg-Landau (theory, parameter, etc.)

FC   in field cooled (specimen)

IS   intermediate state

LMTF   laminar model of the intermediate state in a tilted field

MS   Meissner state

MXS   mixed state

MWM   micro-whirls model

N, n   normal state, phase, etc.

PL   Peierls-London model

RRR   residual resistivity ratio

S, s   superconducting state, phase, etc

ZFC   in zero-field cooled (specimen)

Introduction

Superconductivity is a macroscopic quantum phenomenon with very rich physics and enormous potential for a wide range of applications, many of which would be impossible with other materials. The term macroscopic in this case implies that the radius of microscopic (i.e., non-dissipative) currents determining properties of superconductors is three or more orders of magnitude larger than that in ordinary (non-superconducting or "normal") materials. This and other characteristics of so-called superconducting (i.e., paired) electrons give rise to unusual electric, magnetic and thermal properties of superconductors and provide unique opportunities for testing and clarifying basic concepts of the electrodynamics (the term coined by André-Marie Ampére) of regular media. This includes, for instance, direct measurements of the profile of induction B at an interface of media with different magnetizations, direct measurements of the field strength H controlling the magnetism in matter, and much more. The fundamentals of the electromagnetic and thermal properties of superconducting materials is the subject of this book.

The huge scale of the microscopic currents makes superconductivity a phenomenon immediately adjacent to the realm of the classical physics. Indeed, as the reader will see, the only non-classical formula necessary to describe the properties of superconductors is the Bohr-Sommerfeld quantization condition, written in a generalized form. The rest follows from the classical Faraday-Maxwell electrodynamics [1] and the thermodynamics of magnetic systems of Guggenheim [2, 3]. This fact allows superconductivity to be described in simple mathematical terms with a lucid physical meaning.

Superconductivity was discovered by Kameringh Onnes in 1911 as a phenomenon of a sharp drop of electrical resistance (hence the term supra- or superconductivity ccined by Onnes) in a high-purity mercury wire at a temperature slightly below 4.2 K [4]. An original estimate of the resistance drop was 10^4 times compared to the resistance just above the transition temperature, called the critical temperature T_c, a constant of the superconducting (S) material. Fairly soon it became clear that resistance of a superconductor is not merely a small, but zero.

It was also found that the S state exists up to a certain temperature-dependent critical magnetic field $H_c(T)$, which is another characteristic of the material. While the number of superconductors was (and is) steadily growing, at an early stage it seemed that zero resistivity is the only property distinguishing superconductors from the normal (N) metals. However, further studies have shown that superconductivity is far not only about zero resistance.

The second hallmark of superconductivity, the disappearance of thermoelectric effects, was discovered by Walther Meissner in 1927 [5]. Lack of thermo-e.m.f. suggests that entropy of the S fraction of conduction electrons is zero. If so, then superconductivity can be a previously unknown thermodynamic state of matter characterized by complete ordering of the superconducting electrons. Gorter and Casimir [6] used zero entropy as the base postulate of their two-fluid model (1934), a seminal thermodynamic theory describing properties of superconductors in the absence of a magnetic field.

The idea of the new thermodynamic state was brilliantly confirmed in the Meissner effect independently discovered by Meissner and Ochsenfeld (1933) [7] and Rjabinin and Shubnikov (1934) [8]. It was found that, apart from zero resistivity and zero entropy, superconductivity is also characterized by zero magnetic induction $\mathbf{B}$ regardless on the history of the field application. Hence, a specimen in the Meissner state (MS) represents a perfect diamagnet; i.e., its magnetic permeability μ_m and susceptibility $\chi[= (\mu_n - 1)/4\pi]$ take the minimum possible values zero and $-1/4\pi$, respectively. Important to note that (i) like in regular diamagnetic materials, in the MS χ does not depend on the specimen temperature T and the applied field $\mathbf{H}_0$ and (ii) if the first two "big zeroes" of superconductivity take place in bodies of any connectivity and purity, the third zero ($\mathbf{B} = 0$ or the Meissner effect) is observed only in sufficiently pure massive singly connected bodies of an ellipsoidal shape.

It was found that there are two kinds of superconductors distinguished by the sign of the S/N interphase energy, also called surface tension. Materials with positive and negative surface tension are referred to as type-I and type-II superconductors, respectively. As a rule, type-I superconductors are elementary metals, while type-II materials, with very few exceptions, are alloys and multi-component compounds. A type-I material can be made to behave like a type-II superconductor by introducing impurities and/or reducing the body dimensions, e.g., by decreasing the thickness of a film specimen. However, the opposite is not possible. Other really astonishing properties, such as the flux quantization, Josephson effect and others, were discovered in 1960s and later. The most important of the more recent findings was the revelation of high-temperature superconductors by Müller and Bednorz in 1986 [9]. It turned out that superconductivity is not only the low-temperature phenomenon. In particular, experiments with hydrides at high pressures (see, e.g. [10]) suggest that this phenomenon may not have a fundamental limit on the critical temperature.

A search for the source of the superconductivity culminated in a ground breaking theoretical discovery of Leon Cooper (1956) [11], who has shown that even a weak attraction can lead to formation of stable electron pairs called Cooper pairs. Cooper's prediction was irrefutably confirmed in experiments (Deaver and Fairbank [12], Doll and Näbauer [13], and many others).

Cooper pairs profoundly alter the medium electromagnetic and thermal properties. In fact, the electrodynamics of superconductors is essentially the electrodynamics of an ensemble of Cooper pairs. The electrodynamics of superconductors constitutes the main topic of this book.

At static conditions the role of normal (unpaired) electrons is mostly reduced to screening the external electric field. Their role enhances with increasing frequency ν of the applied field and/or of the transport current. As was shown by Pippard [14] the effect of frequency is minor at $\nu \lesssim 10^9$ Hz. At higher frequencies one needs to take into account skin- and other effects due to unpaired electrons. At $\nu \sim 10^{13}$ Hz properties of superconductors become undistinguished from those of normal metals [15]. On that reason, major properties and applications of superconductors are associated with the static and quasi-static conditions. Accordingly, in this book we will primarily deal with dc magnetic and electric properties.

On the other hand, superconductors are related to magnetizing materials, i.e., to the materials which magnetization vanishes in the absence of an external magnetic field. Therefore, when considering the properties of normal magnetics, materials with spontaneous magnetization are mentioned not often.

The book consists of five chapters. To deepen understanding, the chapters are accompanied by problems.

The first two chapters lay the foundation for understanding superconductivity. In the first chapter we consider general concepts of magnetostatics, such as magnetization $\mathbf{I}$, the magnetic field intensity $\mathbf{H}$ and induction $\mathbf{B}$, their vector potentials $\mathbf{A}$ and $\mathbf{A}_B$, relationships between these quantities and the boundary conditions. The most important sections of this chapter are related to the field $\mathbf{H}$ and its vector potential $\mathbf{A}$, which control magnetic properties of all (superconducting and non-superconducting) media, but, unfortunately, are insufficiently presented in majority of post-Maxwell textbooks.

The second chapter is devoted to gyromagnetic phenomena. The Larmor theorem and historical experiments of Barnett, Einstein and de Haas, and I. Kikoin and Goobar are discussed. Fundamental Langevin's theories of the dia- and paramagnetism are considered as well.

The third chapter concerns historical aspects of the development of superconductivity: the early years in Leiden and the constitutional experiments of Meissner and Ochsenfeld and of Ryabinin and Shubnikov are reviewed. The theories of Gorter and Casimir (the two-fluid model) and the London brothers are examined in detail, theories of Ginzburg and Landau, and of Bardeen, Cooper and Schrieffer are also considered.

Thermodynamics of superconductors is discussed in the fourth chapter. It includes thermodynamic potentials and the key thermodynamic properties distinguishing superconductors from non-superconducting materials, such as the condensation energy and the S/N surface tension. The content of this chapter represents an abridged and updated version of the author's previous book [16], focused on the thermodynamics of magnetizing materials.

The fifth chapter is devoted to proper electrodynamics of superconductors. A theoretical micro-whirls model (MWM) is presented in detail. As the reader will see, the MWM, based on the generalized Bohr-Sommerfeld quantization condition, allows describing all known properties and effects of superconductivity. Some of these properties, which have been acquainted for many decades, are explained for the first time using this model. The MWM suggests a possible existence of yet unknown effects, outlined at the end.

Ancillaries in the Support Material section of the book website of the publisher contain video recordings of the magnetic domains in a type-I superconductor (lead) in the intermediate state, subjected to a dc transport current. The videos were produced by Mariela Menghini and Rinke Wijngaarden in the Vrije Universiteit Amsterdam using the magneto-optical imaging technology.

CHAPTER 1

MAGNETIC FIELDS IN NORMAL MEDIA

1.1   MICROSCOPIC CURRENTS AND MAGNETIZATION

Magnetic properties of any body[1] (we will also use terms "sample" and "specimen", the latter mainly referring to bodies of ellipsoidal shape) are determined by microscopic currents circulating persistently in its interior[2]. An important, although not the only characteristic of these currents, is the magnetic moment (strictly, dipole magnetic moment) of the body, defined as

$$\mathbf{M} = \frac{1}{2c} \int_V (\mathbf{r} \times \mathbf{j}_{mic}) dV, \tag{1.1}$$

where $\mathbf{r}$ is the radius-vector from an arbitrary located origin to a physically infinitesimal volume element[3] of the body dV with the microscopic current of density $\mathbf{j}_{mic} = en\mathbf{v}_{mic}$ with e, n and $\mathbf{v}_{mic}$ are charge, number density and velocity of the genuine microscopic motion of electrons, respectively; the integral is taken over the whole volume of the body V; c is the electrodynamic constant of the *cgs* unit system equal to the speed of light.

In the general case, the microscopic current density of electrons $\mathbf{j}_{mic}$ consists of the molecular current $\mathbf{j}_{mol}$ and the current formed by conduction electrons $\mathbf{j}_c$. In its turn, the latter is divided for microscopic circular current[4]

---

[1] Unless otherwise stated, the body is assumed singly connected.

[2] Recall that the intrinsic magnetic moments of electrons and other elementary particles with non-zero spin angular momentum, are also due to circulating currents [17].

[3] After Lorentz, a physically infinitesimal elements of a volume, area or length are such their elements which are much smaller than macroscopic nonuniformities and at the same time much larger than microscopic inhomogeneities of the medium under question. In the normal (non-superconducting) media the scale of the physically infinitesimal elements is of the order of molecular size; in superconductors it is significantly larger - of the order of the size of Cooper pairs.

[4] Examples are the currents responsible for Pauli paramagnetism and Landau diamagnetism; in superconductors these are predominantly currents in Cooper pairs.

DOI: 10.1201/9781003355786-1

$\mathbf{j}'_c$ and the total current flowing on macroscopic paths[5] $\mathbf{j}_t$; in normal media $\mathbf{j}_t$ dissipates. Thus, the microscopic current density is

$$\mathbf{j}_{mic} = \mathbf{j}_{mol} + \mathbf{j}_c = \mathbf{j}_{mol} + \mathbf{j}'_c + \mathbf{j}_t. \tag{1.2}$$

In Eq. (1.1) $\mathbf{M}$ is the moment due to closed microscopic currents; i.e., it does not include a contribution of the total current.

$\mathbf{M}$ can also be written in terms of a time-mean density of the microscopic currents $\bar{\mathbf{j}}_{mic}$, which we will designate as $\mathbf{j}$. In this case, individual electrons moving along closed microscopic trajectories turn into stationary circular microscopic currents with the corresponding magnetic moments. Then $\mathbf{M}$ can be written as

$$\mathbf{M} = \frac{1}{2c} \int_V (\mathbf{r} \times \mathbf{j}) dV = \int_V \mathbf{I} dV, \tag{1.3}$$

where $\mathbf{I}$ is *magnetization*, the magnetic moment per unit volume determined by the closed microscopic currents. In contrast to the magnetic moment of the body $\mathbf{M}$, magnetization $\mathbf{I}$ is an unambiguous characteristic of all microscopic currents within a body.

In reverse, an averaged density of the microscopic currents $\mathbf{j}$ is expressed through magnetization as [18, 19]

$$\mathbf{j} = c\nabla \times \mathbf{I}. \tag{1.4}$$

In non-conducting materials, referred to as insulators and dielectrics, the conduction component in Eq. (1.2) is absent, i.e., $\mathbf{j}_{mic} = \mathbf{j}_{mol}$.

1.2 DIELECTRICS

1.2.1 Bound current and time reversal symmetry

The time-mean density of molecular current $\bar{\mathbf{j}}_{mol}$ is referred to as the bound current $\mathbf{j}_b$. Respectively, in dielectrics $\mathbf{j}$ in Eqs. (1.3) and (1.4) is replaced by $\mathbf{j}_b$.

The reader probably noticed a strange thing: according to Eq. (1.4) in bodies with uniform (i.e., non-zero) magnetization[6] the current density $\mathbf{j}_b$ is zero and, if so, the magnetic moment and magnetization should be zero as well!

However, this contradiction is only apparent: the current in Eq. (1.4) is the average and it is indeed zero in the specimen bulk due to mutual compensation of the neighboring bound currents. But the magnetic moments of these currents are parallel and summed up, resulting in non-zero magnetization. At the same time near the surface, the bound currents are uncompensated. Then

---

[5]In singly connected bodies $\mathbf{j}_t$ is the density of transport current, also referred to as the conduction and free current. In multiply connected superconductors, the total current also includes currents encircling openings with trapped magnetic flux.

[6]As we will see below, such bodies always have ellipsoidal shape.

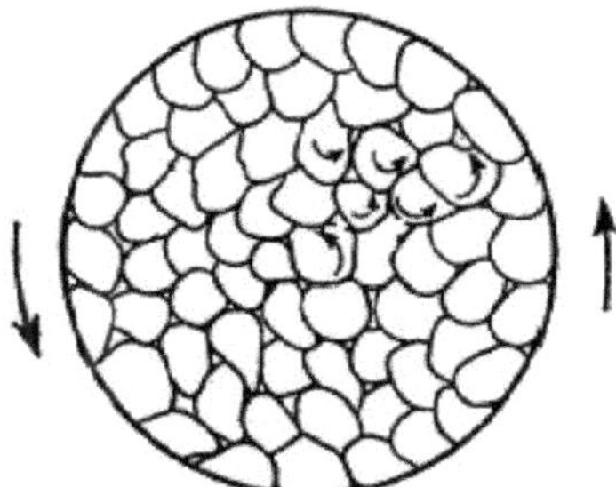

Figure 1.1 A schematic cross section of a cylindrical diamagnetic specimen with bound currents caused by precession of the atomic electron orbits induced by an applied magnetic field; big arrows designate the surface current formed by the bound electrons; the field is directed into the page (After I. Tamm [18] reprinted with permission from Mir Titles).

the specimen magnetic moment is exactly the same as it would be if the total current of the same density as $\mathbf{j}_b$ persistently runs along the surface[7] (see problem 1.1.). This situation is schematically illustrated in Fig. 1.1.

Respectively, Eq. (1.3) in this case can be written as

$$\mathbf{M} = \frac{1}{2c} \oint_S (\mathbf{r} \times \mathbf{g}_b) dS = \int_V \mathbf{I} dV, \tag{1.5}$$

where the first integral is taken over the body surface S, and $\mathbf{g}_b$ is a mean linear density of the surface current; i.e., the bound current near the surface. The latter is defined as

$$\mathbf{g}_b = \int \mathbf{j}_b dl, \tag{1.6}$$

where dl is the length element transverse to the surface.

Through magnetization $\mathbf{g}_b$ is expressed as [18, 19]

$$\mathbf{g}_b = c\mathbf{I} \times \mathbf{n}, \tag{1.7}$$

where $\mathbf{n}$ is a unit vector pointing outward from the body perpendicular to its surface.

The origin of $\mathbf{g}_b$ is a jump in magnetization at the body boundary. Respectively, the depth of the surface current is of the order of radius of the circular microscopic currents near the surface. Hence, although average bound current in the bulk is zero, it is not so near the surface, where this current makes an effective current shell. In words, $\mathbf{g}_b$ is an average current per unit length of a line on the body boundary perpendicular to this current. Let us emphasize that $\mathbf{g}_b$ is formed by closed microscopic currents and does not create the

---

[7] In passing, we note that the sameness of magnetic moments of the solid specimen and its hollow image (current shell) in no means implies the sameness of other electromagnetic properties of these objects.

total current on the specimen surface regardless of whether the specimen is non-conductive, conductive, or superconductive.

On the other hand, the magnetic moment of the current shell in a crucial manner depends on the shape of the shell and therefore on the shape of specimen. Hence, when discussing magnetic properties, the specimen shape has to be specified.

After all, magnetization of a given uniformly magnetized specimen is completely determined by the magnitude and direction of the applied field and, possibly, temperature[8], regardless of their history. This indicates that such a specimen is in a thermodynamic equilibrium state [16]. Therefore, it must obey the Law of symmetry of the time reversal, which implies that all currents in the specimen bulk must mutually compensate each other, so as a total current does not arise[9] [20]. This is precisely what follows from Eqs. (1.4,5,7).

1.2.2 Magnetic induction

Depending on the body shape, magnetization caused by the applied field changes the field both outside and inside the body. The magnetic field in matter is called *magnetic induction* [1], also referred to as the magnetic flux density, denoted **B**. This is a mean value of the magnetic field caused by all microscopic currents or the average of microscopic magnetic fields **h** obeying the Maxwell equations for free space [21]. This means that a probe with non-zero spin (e.g., muon in the μSR spectroscopy) embedded in the specimen will experience the action and, therefore, register the induction $\mathbf{B} \equiv \langle \mathbf{h} \rangle$. The applied field $\mathbf{H}_0$ in this case is the field which would be in the volume occupied by the body if it were not present; for most of the body geometries it is equal to the field far away from the body.

The Maxwell equations for **h** are

$$\nabla \cdot \mathbf{h} = 0 \tag{1.8}$$

and

$$\nabla \times \mathbf{h} = \frac{1}{c}\frac{\partial \mathbf{e}}{\partial t} + \frac{4\pi}{c}\mathbf{j}_{mic}, \tag{1.9}$$

where **e** is a microscopic electric field.

The average electric field in electrically neutral media in the absence of an applied electric field (as assumed in the case under consideration) is zero. Therefore, the time derivative in Eq. (1.9) after its averaging vanishes. Thus, taking into account Eq. (1.4), Eqs. (1.8) and (1.9) upon averaging become

$$\nabla \cdot \mathbf{B} = 0 \tag{1.10}$$

---

[8]As a rule, in diamagnetics **I** does not depend on temperature.

[9]Thermodynamic equilibrium is a state independent of time, while the time transformation $t \to -t$ changes direction of the magnetic moment of a system with total current. Therefore the system (body) with total current cannot be in the equilibrium state by definition.

and

$$\nabla \times \mathbf{B} = 4\pi \nabla \times \mathbf{I}. \tag{1.11}$$

1.2.3 Field intensity and magnetic susceptibility

The fact that both parts of Eq. (1.11) are curls allows one to transform it as

$$\nabla \times \mathbf{H} = 0, \tag{1.12}$$

where $\mathbf{H}$ is

$$\mathbf{H} \equiv \mathbf{B} - 4\pi \mathbf{I}. \tag{1.13}$$

$\mathbf{H}$ is another characteristics of the magnetic field referred to as the *field intensity*. Its other names include the field strength, magnetic field, H-field, auxiliary field, *etc.* Maxwell calls it the magnetizing force. In the free space, where $\mathbf{I} = 0$ by definition, the intensity $\mathbf{H}$ and induction $\mathbf{B}$ are identical and therefore can be used interchangeably. However, inside the matter $\mathbf{H}$ and $\mathbf{B}$ are different and must be *carefully distinguished* [1].

Eq. (1.12) was obtained for the dielectric media. Apparently, it is valid for any other media in the absence of total current. If it is not so, as, e.g., in conductors carrying the total current, the same steps which led us to Eq. (1.12) will yield

$$\nabla \times \mathbf{H} = \frac{4\pi}{c} \mathbf{j}_t. \tag{1.14}$$

This is Ampére's law, which reflects the fact that the field intensity in a medium with the total current (this can be the current in the winding of electromagnet) is determined by this current regardless of magnetic properties of the medium [20, 22]. This means that $\mathbf{H}$ can be interpreted as the field produced by the total current. Below we will see that this is far from the only property of the $\mathbf{H}$ field.

In *cgs* unit system, also referred as Gaussian, absolute and symmetric system, H and B have the same dimension: $[Length]^{-1/2} \times [Mass]^{1/2} \times [Time]^{-1} = cm^{-1/2}g^{1/2}s^{-1}$. After Hans Christian Øersted and Carl Friedrich Gauss, units of H and B are named as the oersted (Oe) and gauss (G), respectively.

In SI units the dimension of B is $[Mass] \times [Time]^{-2} \times [Current]^{-1} = kg \cdot s^{-2} A^{-1}$ whereas the dimension of H is $[Current] \times [Length]^{-1} = A/m$; the unit of B is called the tesla (after Nikola Tesla); there is no name for the unit of H.

As we have seen, in free space $\mathbf{B}$ and $\mathbf{H}$ are identical. Important that this is not just a mathematical equality but a physical reality. The latter with necessity requires identical dimension for B and H. Non-fulfillment of this requirement in the SI is the principal obstacle in using the SI units in electrodynamics of continuous media.

Eq. (1.10) for **B** and Eqs. (1.12) or (1.14) for **H** are basic equations of the field in media. To compose a system they have to be supplemented by a relationship between magnetization **I** and either induction **B** or intensity **H**. This relationship follows from *the fundamental equation of induced magnetism* [1], which has the form

$$\mathbf{I} = \chi \mathbf{H}, \tag{1.15}$$

where χ is the *magnetic susceptibility* or coefficient of induced magnetization per unit volume, a dimensionless quantity characterizing material in question[10].

By definition, Eq. (1.15) is applicable to dia- and paramagnetic media. For ferromagnetic materials it can be used if the field is applied to an initially demagnetized specimen.

Using Eq. (1.15) and the definition of **H** we obtain

$$\mathbf{B} = (1 + 4\pi\chi)\mathbf{H} = \mu_m \mathbf{H}, \tag{1.16}$$

where μ_m is called the magnetic permeability.

In normal materials the permeability is always positive[11]. In superconductors in the Meissner state μ_m reaches its minimal possible value: zero. In not too strong magnetic field the susceptibility χ and, hence, the permeability μ_m are material constants, except superconductors[12]. Materials with a field-independent μ_m are called linear.

In paramagnetic materials, also referred to as paramagnetics and paramagnets, χ is positive and it is negative in diamagnetics or diamagnets. Respectively, in paramagnetics $\mu_m > 1$, while in diamagnetics $\mu_m < 1$. Also, in diamagnetics χ, being different in different substances, depends neither on temperature nor on the field (with some exceptions [23]). In paramagnetics χ decreases with increasing temperature; in many of these materials $\chi \sim 1/T$ (Curie Law).

These general characteristics of dia- and paramagnetics immediately allow one to infer their basic microscopic properties. Specifically, the constancy of χ in diamagnetics indicates that the field-induced microscopic moments are completely ordered and therefore the magnetic part of their entropy $S_m = 0$ (see problem 1.2.). Then, taking into account the absence of spontaneous/residual magnetization, one can conclude that in diamagnetics at zero field the moment

---

[10]In handbooks the susceptibilities can be given per unit mass (the mass susceptibility χ_m) and per mole (the molar susceptibility χ_{mole}). $\chi_m = \chi/\rho$ and $\chi_{mole} = \chi M_{mole}/\rho$, where ρ is the mass density and M_{mole} is the molar mass.

[11]Magnetization (I) cannot change the sign of the average field (B) with respect to the field causing it (H).

[12]The permeability of all superconductor in the Meissner state is zero.

of microscopic constituent units (atoms, molecules, Cooper pairs in a volume element, *etc.*) is zero, implying that the spin and orbital magnetic moments of electrons in these units are compensated. The former, in its turn, implies that the bound electrons in diamagnetics are effectively spinless.

In paramagnetics, the above characteristics along with the absence of spontaneous magnetization indicate the presence of intrinsic magnetic moments of molecules (incomplete compensation of the spin and/or orbital moments of electrons) and a weak (negligible) interaction between the moments. In zero field the moments are totally disordered and therefore compensate each other by virtue of symmetry. The field-induced positive moment arises due to the tendency of the microscopic moments to orient along the acting on them field, i.e., as follows from Eq. (1.15), along the field $\mathbf{H}$. The Curie law indicates that the moments are subject to temperature agitation. Hence, S_m in paramagnetics decreases with the field and increases with temperature. This gives rise to the magnetocaloric effect [24, 16], used, in particular, to achieve very low (in the mK range) temperatures.

According to Eq. (1.15), the intensity $\mathbf{H}$ controls magnetization. On that reason Maxwell calls $\mathbf{H}$ the magnetizing force. One can ask, why it is not the average field in the media $\mathbf{B}$? Historically, the definition Eq. (1.15) appeared in the Poisson theory [25] and remains in forth in the classical Faraday-Maxwell theory [1, 20]. That the field in Eq. (1.15) cannot be $\mathbf{B}$ is obvious from the fact that magnetization is caused by interaction of molecular currents with an average magnetic field acting on a given molecule from its neighbors. These field does not include the field due to molecule itself, while the induction $\mathbf{B}$ does include it.

In literature, e.g.. [18, 22, 23]. one can find different opinions on which field should be used in the definition of χ. The problem is that in normal magnetizing materials the magnetism is fairly weak. In most cases, $\chi \sim 10^{-4}$ in paramagnetics, and even less (but negative) in diamagnetics (see, e.g., [26]). Hence the induction (as mentioned, it can be measured by μSR) and the intensity (which can be measured by the magnetic resonances) have practically indistinguishable values. However, in superconductors (a) in the Meissner state, where $\mathbf{B}$ is zero but $\mathbf{H}$ is not, the concept of the magnetic susceptibility makes sense only if $\mathbf{I}$ is proportional to $\mathbf{H}$; and (b) in superconductors in the intermediate state $\mathbf{H}$ in the superconducting domains is equal to $\mathbf{B}$ in the normal domains, which can be and was measured by μSR [27]. Results of these measurements as well as data on the gyromagnetic effect in superconductors (see Sec. 2.2) confirm the correctness of the classical definition of χ Eq. (1.15).

Thus, the intensity $\mathbf{H}$ is the average field acting on molecules (or other microscopic units composing the magnetizing material) of the specimen. Such units are called "native" (belonging to the specimen). Accordingly, probe charges or currents, such as, e.g., muons in μSR, are called "foreign".

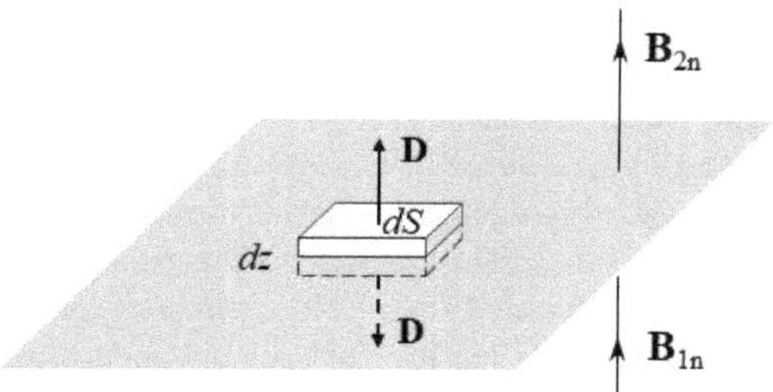

Figure 1.2 To derivation of Eq. (1.17). $\mathbf{B}_{1n}$ and $\mathbf{B}_{2n}$ are components of $\mathbf{B}$ normal to the box surface dS; $\mathbf{D}$ are vectors coming from the box volume $dV = dzdS$; $\mathbf{D} = (\mathbf{B}_{2n} - \mathbf{B}_{1n})/2$.

1.2.4 Boundary conditions for **B** and **H**

In the presence of an interface between media with different magnetization, the system of equations (1.10), (1.12)/(1.14) and (1.16) should be accompanied by the boundary conditions for $\mathbf{B}$ and $\mathbf{H}$ fields.

The boundary condition for the normal components of the induction B_n stems from the always valid Eq. (1.10), reflecting the absence of magnetic monopoles. It reads

$$B_{1n} = B_{2n}, \tag{1.17}$$

where indexes 1 and 2 designate 1st and 2nd medium with magnetization $\mathbf{I}_1$ and $\mathbf{I}_2$, respectively.

To derive Eq. (1.17), imagine a small box with area dS and height dz at the interface between the media, assume that $B_{2n} > B_{1n}$ and introduce a vector $\mathbf{D} = (\mathbf{B}_{2n} - \mathbf{B}_{1n})/2$. Hence, one can say that the change in $\mathbf{B}_n$ occurs due to two vectors $\mathbf{D}$ (not to be confused with the electric induction) coming from the upper and bottom sides of the box and directed opposite to each other as shown in Fig. 1.2. Applying the Gauss theorem, we write

$$\Phi_D = \oint_S \mathbf{D} \cdot d\mathbf{S} = \int_V \nabla \cdot \mathbf{D}\, dV = \nabla \cdot \mathbf{D}\,(dS \cdot dz),$$

where Φ_D is the flux of vector field $\mathbf{D}$ through the total surface area S of the box whose volume $V = dSdz$, and $d\mathbf{S}$ is a vector directed normally to dS outward from the box.

Now, the zero divergence of $\mathbf{B}$ implies the zero divergence of all its components. Therefore, $\nabla \cdot \mathbf{D} = (\nabla \cdot \mathbf{B}_{2n} - \nabla \cdot \mathbf{B}_{1n})/2 = 0$. Hence, $\mathbf{D} = 0$ or $B_{2n} = B_{1n}$. In other words, the lines of induction are always continuous. This means that the B-lines either make closed loops or they come from and leave to infinity[13], or the number of lines entering and exiting some volume is the

---

[13]They can also make an endless toroidal helix [18].

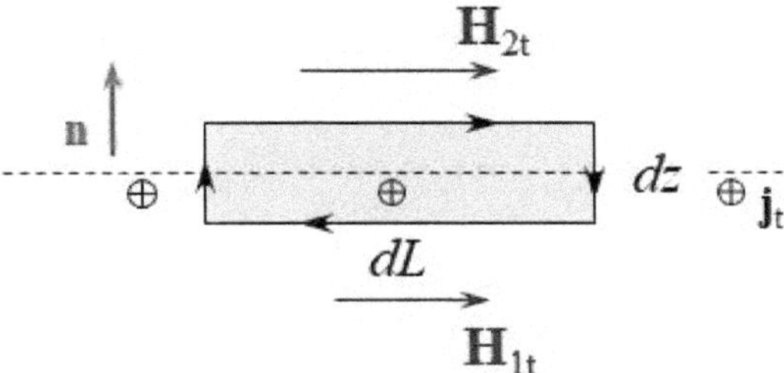

Figure 1.3 To derivation of Eq. (1.18). H_{1t} and H_{2t} are components of $\mathbf{H}$ tangential to the interface between media shown by the dashed line; $\mathbf{n}$ is a unit vector transverse to the interface; $\mathbf{j}_t$ is density of the total current directed into the page.

same. The latter constitutes the law of the magnetic flux conservation, which is thus a consequence of the absence of magnetic monopoles.

The boundary condition for tangential component of the intensity H_t follows from Eq. (1.12). It reads

$$H_{1t} = H_{2t}. \tag{1.18}$$

To derive this condition, consider a cross section of the box in Fig. 1.2 lying in the plane transverse to the interface, as shown in Fig. 1.3. A boundary of the cross section represents a rectangular loop with length dL and width dz. Now, calculate the circulation of the vector $\mathbf{H}$ along the loop in the limit $dz \to 0$ while keeping dL fixed; take the path direction shown by arrows as a positive.

If $\mathbf{j}_t = 0$, then using the Stocks theorem[14] and Eq. (1.12), we obtain

$$\oint_l H_l dl = (H_{2t} - H_{1t})dL = \int_a (\nabla \times \mathbf{H}) \cdot d\mathbf{a} = 0, \tag{1.19}$$

where l is the total length of our rectangular loop, dl is the path element, a is the loop area equal to $dLdz$, $d\mathbf{a}$ is the vector element of the area directed into the page and H_l is a projection of $\mathbf{H}$ on the path.

Thus, we have proven that in the absence of total current the tangential component of the intensity is continuous at the interface between media. Important that this is true for any contacting media: regardless whether both or one of them is diamagnetic, paramagnetic or ferromagnetic, or whether they are dielectrics, conductors or superconductors.

But what will change if $j_t \neq 0$? For most cases the answer is nothing. Indeed, by definition. the total current $J_t = \mathbf{j}_t \cdot \mathbf{a}$. Since when $dz \to 0$ the area $|\mathbf{a}|$ tends to zero as well, Eq. (1.18) remains the same providing $\mathbf{j}_t$ is finite, as it often takes place. However, if the total current with a sharply distributed

---

[14]About the name of this theorem, see the interesting paper by Katz [28].

(as in a δ-function) density is concentrated in a thin layer near the boundary, Eq. (1.18) takes the form (see problem 1.3)

$$(\mathbf{H}_1 - \mathbf{H}_2) \times \mathbf{n} = \frac{4\pi}{c}\mathbf{g}_t, \qquad (1.20)$$

where $\mathbf{g}_t$ is the linear density of the *total* (macroscopic) current on the interface and $\mathbf{n}$ is the unit vector transverse to the interface and directed toward medium 2.

In the scalar form Eq. (1.20) reads

$$H_{1t} - H_{2t} = \frac{4\pi}{c}g_t. \qquad (1.21)$$

Thus, at the interface between the media, the tangential component of the intensity H_t can change only in the presence of a total current, provided that the latter flows in the immediate vicinity of the interface. The bound current, as well as a continuously distributed total current (e.g., a current that exponentially decays with depth from the interface), cannot change H_t.

Let us return to the non-conducting media and calculate the boundary conditions for remaining components of $\mathbf{B}$ and $\mathbf{H}$.

Eq. (1.7) is written for the interface of the magnetized medium and vacuum. In case of two media with magnetizations $\mathbf{I}_1$ and $\mathbf{I}_2$, this equation becomes

$$\mathbf{g}_b = c(\mathbf{I}_1 - \mathbf{I}_2) \times \mathbf{n}, \qquad (1.22)$$

where the unit vector $\mathbf{n}$ is directed toward the second medium as shown in Fig. 1.3.

By definition, $\mathbf{g}_b$ is directed tangentially to the interface. Therefore, the scalar form of Eq. (1.22) is

$$\frac{4\pi}{c}g_b = (B_{1t} - H_{1t}) - (B_{2t} - H_{2t}). \qquad (1.23)$$

Hence, using the continuity of H_t (Eq. (1.18)), we arrive at the boundary condition for the tangential components of the induction

$$B_{1t} - B_{2t} = \frac{4\pi}{c}g_b. \qquad (1.24)$$

In the vector form this is

$$\frac{4\pi}{c}\mathbf{g}_b = (\mathbf{B}_1 - \mathbf{B}_2) \times \mathbf{n} = 4\pi(\mathbf{I}_1 - \mathbf{I}_2) \times \mathbf{n}. \qquad (1.25)$$

Hence, contrarily to the continuity of the normal component of induction, its tangential component experiences a jump caused by the difference in magnetization and corresponding surface current formed by the bound electrons. Evidently that the jump in B_t does not infringe the continuity of the magnetic induction lines.

We are left to calculate the boundary condition for the normal component of the intensity H_n. Using definition of $\mathbf{H}$ (Eq. (1.13)) and the fact of continuity of B_n, we obtain

$$H_{1n} - H_{2n} = 4\pi(I_{2n} - I_{1n}). \tag{1.26}$$

This condition is similar to that for the electrostatic field $\mathbf{E}$ in metals and polarized dielectrics $E_{1n} - E_{2n} = 4\pi\sigma$, where σ is a surface density of the free and bound charges, respectively, serving as sources and sinks of the $\mathbf{E}$-lines. Hence, in the time-independent magnetic field I_n at the specimen surface plays the role of σ in electrostatics and therefore the lines of the intensity H at the surface of the magnetized specimen undergo a break, exactly as it takes place in the theory of the magnetic charges [25, 1, 18]. This property provides one more interpretation of the H-field: it can be considered as a potential field created by the "surface charges" with density $(I_{2n} - I_{1n})$. We will return to this point a little later.

1.2.5 Demagnetizing field and relationship between $\mathbf{B}$, $\mathbf{H}$ and $\mathbf{H_0}$

Eqs. (1.10), (1.12)/(1.14) and (1.16) with boundary conditions (1.17), (1.18), (1.24) and (1.26) represent a complete system of equations, which in principle allows one to calculate all magnetic properties of any magnetized body. However, practical use of this approach to the bodies of an arbitrary shape is limited by enormous complexity of the structure of bound currents due to the inhomogeneous magnetization in such bodies. Fortunately, there is a large group of bodies that allow rigorous analytical calculations of their magnetic properties. Those are the bodies of ellipsoidal shape, which we agreed to call specimens.

As it was shown by Poisson [25] (see also [1]), in a body bound by a complete surface of a second degree (i.e., in an ellipsoid) subjected to a uniform field $\mathbf{H_0}$ the intensity $\mathbf{H}$ is uniform although it can differ from $\mathbf{H_0}$ both in magnitude and direction. The same is true for the electric field $\mathbf{E}$ in polarized dielectric ellipsoids [20]. This Poisson theorem is applicable to specimens of any material, including superconducting and ferromagnetic ones, provided the latter were not magnetized prior H_0 is applied. Moreover, due to the continuity of H_t, for superconductors the Poisson theorem is valid for the specimens with both homogeneous and inhomogeneous magnetization[15]. For homogeneous specimens the Poisson theorem allows one to link analytically $\mathbf{B}$ and $\mathbf{H}$ with the applied field $\mathbf{H_0}$ that caused them.

To remind, the ellipsoid is a figure characterized by three mutually perpendicular axes a, b and c, and its volume is $V = 4\pi abc/3$. The difference between $\mathbf{H_0}$ and $\mathbf{H}$ inside the ellipsoid is referred to as *demagnetizing field* $\mathbf{H_d}$. It is proportional to magnetization $\mathbf{I}$ and depends on the specimen geometry, i.e., on

---

[15]On this reason equilibrium superconducting states, such as the Meissner, intermediate and mixed states, are observed only in the ellipsoidal bodies.

relationship between the ellipsoidal axes. If an orthogonal coordinate system is oriented so that its axes (x,y,z) are parallel to the axes (a,b,c), respectively, then in the homogeneously magnetized specimens the x-component of $\mathbf{H}_d$ is

$$(H_d)_x = (H_0)_x - H_x = 4\pi I_x \eta_x, \qquad (1.27)$$

where η_x is called *demagnetizing factor* with respect to that $(x\text{-})$ axis. η_x is a coefficient of proportionality between $(H_d)_x$ and I_x, and 4π is introduced for consistency with other formulae in *cgs* unit system. Replacing subscript x by y and z, the Eq. (1.27) is valid for the y- and z-axes, respectively. The demagnetizing factors η_x, η_y and η_z are positive coefficients depending *only* on the shape of ellipsoid [29], and their sum meets the identity [1, 20]

$$\eta_x + \eta_y + \eta_z = 1. \qquad (1.28)$$

The simplest example of an ellipsoid is a sphere, for which $\eta_x = \eta_y = \eta_z = 1/3$ due to symmetry. Another example is a long circular cylinder, representing a strongly prolate ellipsoid of revolution. In the limit of $a \gg b = c$ the demagnetizing factor for the longitudinal axis is zero and that for two other axes (transverse to the first one) are $1/2$. If a base of the long cylinder is oval so that $a \gg b > c$, then $\eta_a \to 0$ and $\eta_b \to (1 - \eta_c) < \eta_c$, where η_a, η_b and η_c are the demagnetizing factors with respect to coordinate axes parallel to corresponding axes of the ellipsoid. After all, if both a and b are much greater than c ($a \sim b \gg c$), then η_a and η_b are close to zero whereas $\eta_c \to 1$. In this limit the base of the cylinder represents a very long and narrow rectangle and the entire ellipsoid represents an infinite parallel-plane plate or slab (implying that the plate lateral dimensions greatly exceed its thickness). Evidently that the infinite plate can be also viewed as a strongly oblate ellipsoid, which demagnetizing factor with respect to the axis perpendicular to the plate $\eta_\perp = 1$ and the factors with respect to any axis parallel to the plate $\eta_\parallel = 0$. For illustration, cross-sectional views of specimens with different demagnetizing factors are shown in Fig. 1.4. Detailed graphs and tables allowing to determine η for any axis of an ellipsoid of any shape were composed by Osborn [29].

If the applied field $\mathbf{H}_0$ is parallel to one of the ellipsoidal axes with respect to which the demagnetizing factor is η, then $\mathbf{H}$ and $\mathbf{B}$ inside are parallel to $\mathbf{H}_0$ and related as

$$(1 - \eta)H + \eta B = H_0. \qquad (1.29)$$

If $\mathbf{H}_0$ is not parallel to either of ellipsoidal axes, it should be broken for components parallel to the axes and the components of $\mathbf{B}$ and $\mathbf{H}$ are computed independently.

Replacing H, B and H_0 by E, D (electric induction or displacement) and E_0, respectively, the same equation is valid for the polarized dielectric ellipsoid in an electrostatic field E_0; η in this case is a depolarization factor.

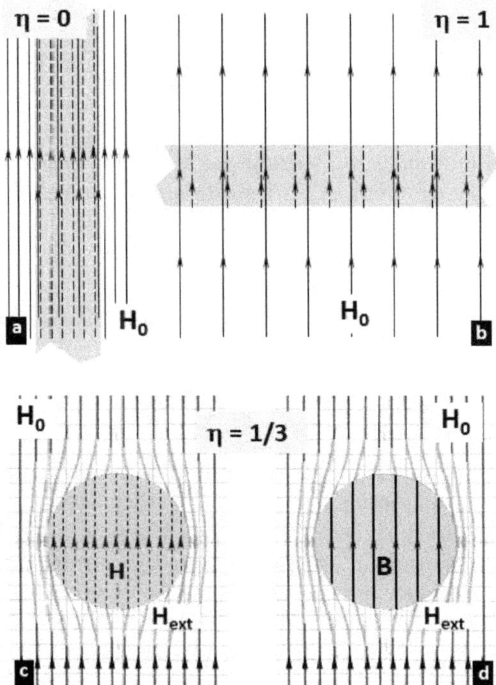

Figure 1.4 Induction **B** (solid lines) and intensity **H** (dashed lines) of a magnetic field inside and outside diamagnetic specimens (cross-section shown in gray) with different demagnetizing factors η. The specimens are in free space. Outside the specimens **B** and **H** are identical. (a) $\eta = 0$, the specimen of a cylindrical geometry (e.g., a long cylinder in a parallel field); (b) $\eta = 1$, the specimen of a transverse geometry (an infinite plate in the perpendicular field); (c) and (d) $\eta = 1/3$, a sphere. $\mathbf{H}_0$ is the applied field; $\mathbf{H}_{ext}$ is the outer field near the specimen. The line densities correspond to the specimens with $\mu_m \approx 0.8$ in (a) and (b), and with $\mu_m \approx 0.5$ in (c) and (d). In all cases, inside the specimens the **H**-field is uniform; **B**-field is uniform either since μ_m is constant. Note that in diamagnetics **B** is lesser than $\mathbf{H}_0$, whereas **H** is greater than $\mathbf{H}_0$. (Reprinted from [16].)

A heuristic derivation of Eq. (1.29) is available in [16]. Maxwell [1] derived it using the scalar potential of the field intensity. Landau and Lifshitz [20] derived Eq. (1.29) for the polarized dielectric ellipsoid. An example of using this equation is available in Problem 1.4.

If either B, or H, or relationship between them are known, Eq. (1.29) along with definition (1.13) allows one to calculate **B**, **H** and **M** for uniformly magnetized ellipsoidal bodies. In particular, the magnetic moment of a specimen made of material with given either μ_m or χ is

$$M = -\frac{V}{4\pi}\frac{1 - \mu_m}{1 - \eta(1 - \mu_m)}H_0 = \frac{\chi}{1 + 4\pi\eta\chi}V\mathbf{H}_0. \tag{1.30}$$

We see that the specimen magnetic properties depend on its material (μ_m) *and* on the specimen shape and orientation of the applied field (η). For instance, the magnetic moment of an infinite slab in parallel field ($\eta = 0$) is

$$M = -\frac{V}{4\pi}(1 - \mu_m)H_0. \tag{1.31}$$

Whereas the moment of the same slab in the perpendicular field ($\eta = 1$) is

$$M = -\frac{V}{4\pi}\frac{(1 - \mu_m)}{\mu_m}H_0. \tag{1.32}$$

Thus, M of paramagnetic specimens is always positive, whereas it is always negative for diamagnetics. Naturally, for normal dia- and paramagnetics, in which μ_m only barely differs from unity, there is almost no difference in magnitude of M measured in field of different orientations; the same is also true for H and B. However, magnetic properties of superconducting bodies, in which μ_m can be much less than unity (down to zero in the Meissner state), as well as properties of bodies made of ferromagnetic materials ($\mu_m \gg 1$) strongly depend on the specimen geometry and the field orientation.

Note, that according to Eq. (1.32) M of a homogeneous superconducting plate (the plate with $\mu_m = 0$ over the volume) in perpendicular field is infinite, and therefore magnetic energy of such a specimen ($E_m = -\mathbf{M}\cdot\mathbf{H}/2$) is infinite as well, which clearly contradicts the law of energy conservation. On that ground Maxwell concluded that materials with $\mu_m = 0$ do not exist [1]. In reality, however, this means that superconducting plate in the perpendicular field can never be in the Meissner state, in full consistency with experiment (see Ch. 4). Superconducting plates in a transverse field can be found only in an inhomogeneous state (i.e., in the intermediate state for type-I and the mixed state for type-II superconductors, respectively) no matter how small the field is.

1.2.6   Vector potential of **B** and scalar potential of **H**

The Helmholtz theorem of the vector calculus (see, e.g., [19]), states that any vector field $\mathbf{F}(\mathbf{r})$ can be presented as the sum of the taken with negative sign gradient of some function of coordinates $\Psi(\mathbf{r})$ and the curl of another coordinate function $\mathbf{A}(\mathbf{r})$. Ψ and $\mathbf{A}$ are referred to as scalar and vector potentials of this field, respectively. Coming from that one can show that if $\nabla \times \mathbf{F} = 0$, then such a field is completely determined by the scalar potential, namely $\mathbf{F} = -\nabla\Psi$. And other way around, if $\nabla \cdot \mathbf{F} = 0$, then the field is characterized by the vector potential as $\mathbf{F} = \nabla \times \mathbf{A}$. The divergence-less and curl-less

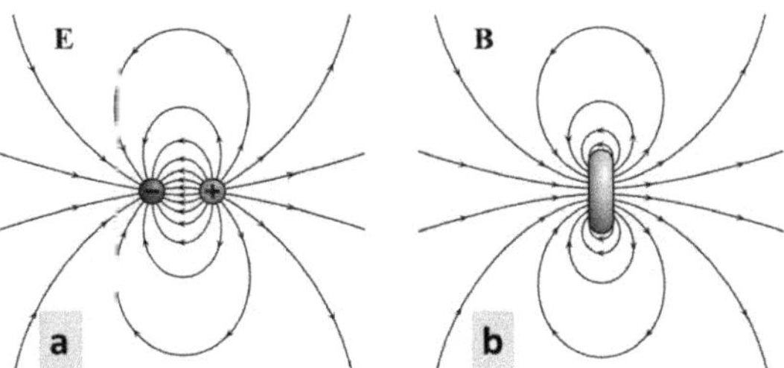

Figure 1.5 The field lines due to electric (a) and magnetic (b) dipoles. The lines of the electrostatic field $\mathbf{E}$ are non-closed; they start on the positive charge (the source) and end on the negative charge (sink); one end (but not both) of these lines can come from or leave to infinity. The lines of the magnetic field (one can equally call it $\mathbf{B}$ or $\mathbf{H}$) are closed or come from and leave to infinity.

fields are called solenoidal (or vortex) and potential (or conservative) fields, respectively [18].

Translated into physics, a potential field is a field created by stationary charges, and a solenoidal field is a field due to moving charges, i.e., currents. In other words, the potential field always has a source and sink (or just the source), whereas the solenoidal field has neither. Well known examples of the potential and solenoidal fields are the electrostatic and magnetic fields, respectively. The scalar potential of the electrostatic field and three components of the vector potential of the magnetic field form 4-vector of the electromagnetic field in the relativity theory [31]. Schematic patterns of the fields generated by the electric and magnetic dipoles are shown in Fig. 1.5.

According to Eq. (1.10), $\mathbf{B}$ is always a solenoidal field. As mentioned, this fact indicates the absence of magnetic monopoles in nature (see, e.g., [17]). Therefore, $\mathbf{B}$ possesses a vector potential $\mathbf{A}_B$[16] defined so that [1]

$$\nabla \times \mathbf{A}_B = \mathbf{B}. \tag{1.33}$$

Usually, the vector potentials are chosen as divergenceless vectors, referred to as Coulomb gauge [20]. For $\mathbf{A}_B$ this means

$$\nabla \cdot \mathbf{A}_B = 0. \tag{1.34}$$

For a system of closed localized currents, e.g., for the currents in a magnetized specimen, the condition of the Coulomb gauge Eq. (1.34) is equivalent to the requirement that all currents remain within the system; i.e., the normal component $\mathbf{j}_n$ at the specimen surface is zero (see [18] for proof). This means that for isolated bodies the Coulomb gauge of the vector potential is presumed automatically.

---

[16] We use $\mathbf{A}_B$ instead of traditional $\mathbf{A}$ because the latter is reserved for the vector potential of intensity. In free space $\mathbf{A}_B = \mathbf{A}$.

A general expression of the vector potential due to currents confined in a closed space, e.g., due to microscopic currents in a specimen of volume V is

$$\mathbf{A}_B = \frac{1}{c} \int_V \frac{\mathbf{j}_{mic}}{R} dV, \tag{1.35}$$

where R is a distance from a current element $\mathbf{j}_{mic}dV$ to a point of observation, which is assumed locating in free space (see problem 1.5).

Outside the specimen (i.e., at distances much greater than the radius of the circulating microscopic currents) the field and therefore its vector potential is determined by the magnetic moment $\mathbf{M}$ due to these currents. The expression for $\mathbf{A}_B$ in terms of $\mathbf{M}$ is (see, e.g., [18, 20] for derivation)

$$\mathbf{A}_B = \frac{\mathbf{M} \times \mathbf{R}}{R^3}, \tag{1.36}$$

where $\mathbf{R}$ is a radius vector with the origin in an arbitrary point inside the specimen to a point of observation outside it.

Accordingly, outside the specimen the induction $\mathbf{B}$ (and the intensity $\mathbf{H} = \mathbf{B}$) caused by the magnetized specimen is (see problem 1.6)

$$\mathbf{B} = \nabla \times \mathbf{A}_B = \frac{3(\mathbf{M} \cdot \mathbf{R})\mathbf{R}}{R^5} - \frac{\mathbf{M}}{R^3}. \tag{1.37}$$

One can notice a formal identity of the induction $\mathbf{B}$ in this formula with the electric field $\mathbf{E}$ outside a polarized dielectric specimen with an electric dipole moment $\mathbf{P}$, which is [18]

$$\mathbf{E} = -\nabla\Psi_E = \frac{3(\mathbf{P} \cdot \mathbf{R})\mathbf{R}}{R^5} - \frac{\mathbf{P}}{R^3},$$

where $\Psi_E = \mathbf{P} \cdot \mathbf{R}/R^3$ is a potential of an electrostatic field caused by the dipole.

This implies that the magnetic field in the current free space can be described by a scalar potential Ψ_M equal to (problem 1.7)

$$\Psi_M = \frac{\mathbf{M} \cdot \mathbf{R}}{R^3}, \tag{1.38}$$

which obeys the Laplace equation

$$\nabla^2\Psi_M = 0. \tag{1.39}$$

As one can see from Fig. 1.5, the possibility of using the scalar potential for the magnetic field stems from similarity of the electric and magnetic fields caused by, respectively, electric and magnetic dipoles away from them. This circumstance sometime is interpreted as though the magnetic field in the current-free space can be considered as a potential field with all the ensuing consequences. This includes impossibility of voids and extremums within the field (Earnshaw's theorem). However, it should not be forgotten that the sameness of the magnetic and electric fields is purely formal: contrarily to the potential (curl-free) electrostatic field, the magnetic field is *always* solenoidal (divergence-free). This means that a physically adequate description of the magnetic field requires the use of a vector rather than a scalar potential. In particular, the Laplace equation can be inapplicable in a current-free space near superconductors (see Sec. 4.3.2).

The formal applicability of the scalar potential for the magnetic field can also be justified in another way. As it was mentioned above, the discontinuity of the normal component of intensity H_n at the specimen boundary (Eq. (1.26)) allows one considering the normal component of magnetization at the boundary I_n as the surface density of "magnetic charges" like the surface density of the genuine bound electric charges of the polarized dielectric specimen of the same geometry as the magnetized specimen. This means that formulae of electrostatic field $\mathbf{E}$, including the formula for the scalar potential Ψ_E can be applied to the magnetic field $\mathbf{H}$ (equal to $\mathbf{B}$ outside the specimen), and therefore the pattern of the $\mathbf{E}$ and $\mathbf{H}$ lines should be the same.

We figured out the field outside the magnetized specimen, where $\mathbf{B} = \mathbf{H}$. The next question is how to find $\mathbf{B}$ and $\mathbf{H}$ inside the specimen, where they differ. In principle, one can first calculate $\mathbf{A}_B$ using Eq. (1.35), then take the curl to compute $\mathbf{B}$. After that, if μ_m is known, one can find $\mathbf{H} = \mathbf{B}/\mu_m$. However, this strategy is hardly feasible due to complexity of the structure of microscopic currents even in uniformly magnetized specimens. Moreover, for superconductors such an approach is meaningless inasmuch $\mu_m = 0$ inside them.

However, this problem is easily resolved first computing $\mathbf{H}$ (e.g., using Eq. (1.27)) and then $\mathbf{B} = \mu_m \mathbf{H}$. So, the ability to treat $\mathbf{H}$ as a potential field really helps to solve seemingly unsolvable problems.

As mentioned, this ability is based on the discontinuity of $\mathbf{H}_n$ and therefore non-zero divergence of $\mathbf{H}$ on the specimen surface. This fact mimics an appearance of magnetic charges on the surface, but of course does not create them. Due to that from the physics viewpoint the value of Ψ_M is limited. We have already mentioned an issue with using the Laplace equation in the free space outside a superconducting specimen. Similar issues may happen at consideration of inner properties. For instance, the potential interpretation of intensity cannot explain the gyromagnetic effect as well as phenomena of the magnetic resonance that underlie the most powerful tools for studies of microscopic properties of condensed matter.

The absence of magnetic monopoles unambiguously means that a physically adequate discussion of magnetic properties should be based on the vector potential. On the other hand, due to composite nature of the intensity (Eq. (1.13)), the specimen boundary should be excluded from this consideration. Indeed, in any dV not crossing the boundary, whether it is outside or inside the specimen, $\nabla \cdot \mathbf{H} = 0$. This allows one to introduce the vector potentials for the $\mathbf{H}$-field on the both sides from the boundary. At the same time, due to the change in magnetization, the vector potentials of **H** inside and outside the specimen are different.

1.2.7   Vector potential of **H**

The vector potential **A** of the intensity **H** is defined so that

$$\mathbf{H} = \nabla \times \mathbf{A} \tag{1.40}$$

with the Coulomb gauge presumed automatically.

To find **A** let us consider a mono-atomic non-conducting diamagnetic specimen. All electrons in this specimen are bound to atoms. At zero field, as we already understood (Sec. 1.2.3), the orbital and spin magnetic moments of electrons in each atom are mutually compensated. This means that electrons are effectively spinless and the sum of orbital magnetic moments of electrons in each atom is zero.

When $\mathbf{H}_0$ is turned on, the intensity inside the specimen changes from zero to **H** and each electron experiences an action of the Lorentz force $\mathbf{F}_L$ in addition to the forces acting on it at zero field. In the absence of an applied electric field the Lorentz force is

$$\mathbf{F}_L = -\frac{e}{c}\left(\frac{\partial \mathbf{A}}{\partial t}\right) + \frac{e}{c}\mathbf{v} \times \mathbf{H}, \tag{1.41}$$

where **A** is the vector potential of **H**, **v** is the electron velocity and t is time.

The action of the Lorentz force is considered in detail in Sec. 2.1. Here we just note that the second term of the right hand side is the magnetic force $\mathbf{F}_M$, which makes the electron's orbit (or the magnetic moment $\boldsymbol{\mu}$ due to the orbiting electron) precessing[17] about the field **H** with the Larmor frequency o equal to $-\gamma\mathbf{H}$, where $\gamma(= e/2mc$ with m equal to the free electron mass) is a gyromagnetic ratio of orbiting spinless electrons. Herewith electrons in each atom precess synchronously, i.e., without changing their mutual orientation, and parameters of the orbits (the orbital radius R and the magnetic moment μ in zero field) are unaffected by $\mathbf{F}_M$. The Larmor precession caused by $\mathbf{F}_M$ for a single orbiting electron is schematically shown in Fig. 1.6. Overall, the force $\mathbf{F}_M$ does not change the state of diamagnetic specimen because it does not do work. In other words, $\mathbf{F}_M$ *cannot* create magnetism.

The magnetization of diamagnets is entirely determined by the first term of the Lorentz force. This is the electric force $\mathbf{F}_E$ caused by the induced vortex electric field existing while the magnetic field is changing. As shown in

---

[17]By definition, precession is the periodic motion of a spinning system occurring under the action of a torque directed perpendicular to the axis of spinning. In the given case the spinning system is the current due to orbiting electron, which angular momentum ι is perpendicular to the orbit plane and the magnetic moment $\boldsymbol{\mu}$ is directed in opposite direction. The force $\mathbf{F}_M$ generates a torque $\boldsymbol{\tau} = \boldsymbol{\mu} \times \mathbf{H}$. Thus, $\boldsymbol{\tau}$ is perpendicular to ι and therefore $\boldsymbol{\mu}$ precesses.

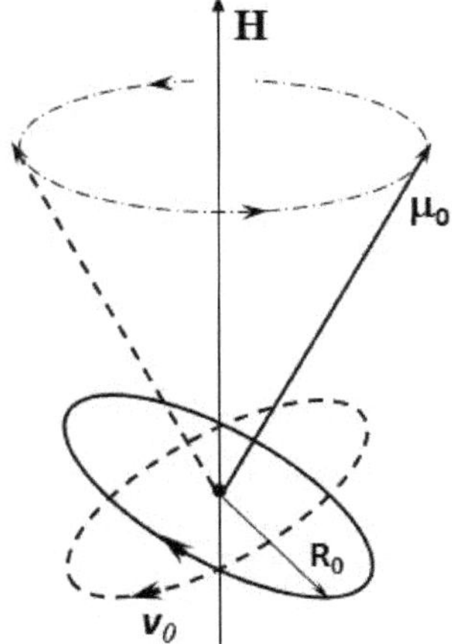

Figure 1.6 Precession of a magnetic moment $\boldsymbol{\mu}_0$ of an orbiting electron caused by the magnetic part $\mathbf{F}_M$ of the Lorentz force. The solid and dashed circles are electron orbits at times separated by half the precession period. v_0, R_0 and $\boldsymbol{\mu}_0$ are velocity, the orbit radius and the magnetic moment of the orbiting electron at zero field, respectively. The dot in the orbit center is an electron-nucleus center of inertia coinciding with the nucleus. The upper dash-dotted circumference depicts the trajectory of the tip of $\boldsymbol{\mu}_0$; the vector $\boldsymbol{\mu}_0$ describes a cone around $\mathbf{H}$. The angular momentum due to the orbiting electron ι_0 (not shown) is directed opposite to $\boldsymbol{\mu}_0$; in cgs units $\iota_0 = 1.14 \cdot 10^{-7} \mu_0$ $(g\,cm^2/s)$.

Sec. 2.1, in a result of its action each orbiting electron acquires an additional field-induced velocity $\mathbf{v}_i$ equal to

$$\mathbf{v}_i = -\frac{e}{cm}\mathbf{A}. \tag{1.42}$$

This is the formula of interest to us in this section.

It shows that $\mathbf{v}_i$ is parallel to $\mathbf{A}$. Therefore, since $\mathbf{A}$, by definition (Eq. (1.40)), lies in the plane perpendicular to $\mathbf{H}$, $\mathbf{v}_i$ and, consequently, the induced current J_i also lie in this plane. On the other hand, the induced current makes a closed loop (it cannot leave the specimen) and the loop must be circular by virtue of rotational symmetry stemming from the uniformity of $\mathbf{H}$ (all directions in planes transverse to $\mathbf{H}$ are equivalent); therefore lines of the vector potential make the circular loops either. Hence, $\mathbf{A}$ should be taken in the circular gauge, namely

$$\mathbf{A} = \frac{\mathbf{H} \times \mathbf{r}}{2}, \tag{1.43}$$

where $\mathbf{r}$ is a radius-vector with the origin at the nucleus, lying in the plane transverse to $\mathbf{H}$. The lines of this vector potential are shown in Fig. 1.7.

Further, since $e < 0$, the lines of the vector potential in Fig. 1.7 also represent the lines of the vector $\mathbf{v}_i$, i.e., trajectories of the induced motion of

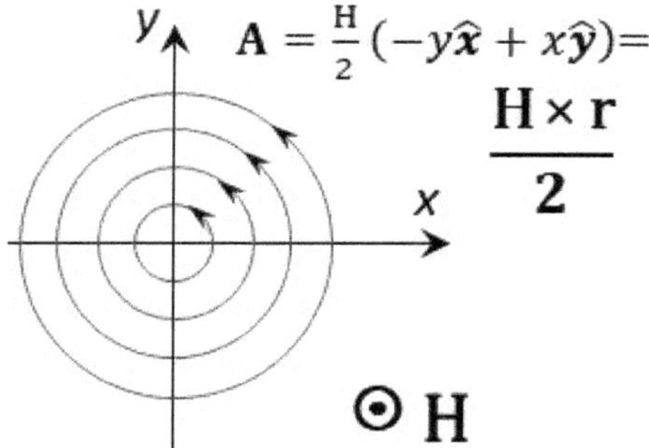

Figure 1.7 Lines of the vector potentials $\mathbf{A}$ in the circular gauge for a uniform magnetic field $\mathbf{H}$ directed toward the reader (along the z-axis); $\hat{x}$ and $\hat{y}$ are unit vectors along x and y axes, respectively. $\mathbf{r}$ is a radius vector lying in xy plane transverse to $\mathbf{H}$. As for all vector fields, the line of $\mathbf{A}$ is the directional line tangential to the vector $\mathbf{A}$ in each of its point.

electrons. The magnetic moment due to this motion is negative and its angular frequency is

$$\boldsymbol{\omega}_i \equiv \frac{\mathbf{r} \times \mathbf{v}_i}{r^2} = -\frac{e}{cm}\frac{\mathbf{r} \times \mathbf{A}}{r^2} = -\frac{e}{2cm}\mathbf{H}, \tag{1.44}$$

where we used Eqs. (1.42) and (1.43), and the "back of the cab" vector identity[18].

Hence, $\boldsymbol{\omega}_i$ is parallel to $\mathbf{H}$ and equal to the Larmor frequency $\mathbf{o}$. Therefore, $\mathbf{A}$ of the circular gauge really provides the Larmor precession. Recall that the regular effect of a magnetic field on a moving charge is a circular motion with the so-called cyclotron frequency $\omega_c = (e/mc)H$.

The last question, which we have to address in this section concerns the relationship between $\mathbf{A}$ and $\mathbf{A}_B$. By definition of the magnetic permeability μ_m, it is

$$\mathbf{A}_B = \mu_m \mathbf{A}. \tag{1.45}$$

1.2.8   On gauge invariance

The scalar and vector potentials as defined in the Helmholtz theorem are ambiguous functions. In particular, the vector potential is determined within the gradient of an arbitrary function of coordinates. The condition of the Coulomb gauge restricts this ambiguity, but does not eliminate it. For example, the vector potential of a uniform solenoidal field $\mathbf{Z}$ directed along the z-axis can be taken in the circular gauge Eq. (1.43), or in two other gauges

$$\mathbf{A}_Z = -Zy\hat{x} \tag{1.46}$$

---

[18]$\mathbf{a} \times (\mathbf{b} \times \mathbf{c}) = \mathbf{b}(\mathbf{a} \cdot \mathbf{c}) - \mathbf{c}(\mathbf{a} \cdot \mathbf{b})$

and

$$\mathbf{A}_Z = Zx\hat{\mathbf{y}}, \tag{1.47}$$

where $\hat{\mathbf{x}}$ and $\hat{\mathbf{y}}$ are unit vectors in directions of x and y axes, respectively (see problem 1.9.).

The independence of a solenoidal field on the choice of vector potential within a definite class of functions (gauges) is called gauge invariance. Since the magnetic field is solenoidal and because in many cases its action is determined by the curl of the vector potential, the gauge invariance for the magnetic field is often considered as a general principle analogous to the principle of equal dimensions underlying dimensional analysis.

The gauge invariance makes an impression that the vector potential is a kind of a mathematical instrument not carrying a specific physical meaning. However, if a physical quantity is determined by the vector potential in the explicit form (i.e., not by its curl), then the notion of gauge invariance is inapplicable [32]. It turned out that such quantities indeed exist and not uncommon. An example is the phase of an electromagnetic wave, which can be changed by the vector potential even in areas where $\mathbf{B} = 0$. This phenomenon is known as Ehrenberg-Siday-Aharonov-Bohm effect [33, 34].

Another, much more common quantity, directly determined by the vector potential, is magnetization. Indeed, as we have seen, the vector potential $\mathbf{A}$ of the intensity $\mathbf{H}$ must meet the rotational symmetry, and the only gauge of such kind is the circular one. Mathematically, the inapplicability of the gauge invariance to magnetization is obvious either from Eqs. (1.42) and (1.44), or from a formula for the induced magnetic moment, which is (see Sec. 2.1)

$$\boldsymbol{\mu}_i = \frac{r^2 e}{2c}\omega_i = \frac{r^2 e}{2cr^2}(\mathbf{r} \times \mathbf{v}_i) = -\frac{e^2}{mc^2}(\mathbf{r} \times \mathbf{A}). \tag{1.48}$$

These facts make clear that the vector potential is a real and primary physical characteristics of the magnetic field. Specifically, if the scalar potential determines the change of the potential energy of a charge in the electrostatic field, the vector potential determines the change of the charge kinetic energy caused by the magnetic field.

The gauge invariance in standard theories of superconductivity. Leaping ahead, it is appropriate to make here a brief stop at a question on the gauge invariance in standard superconductivity theories. Those are the most popular theories discussed in the third chapter: the two-fluid model of Gorter and Casimir, the London theory, the Ginzburg-Landau (GL) theory and the theory of Bardeen, Cooper and Schrieffer (BCS). The last three theories consider properties of superconductors in a magnetic field. Are these theories gauge-invariant?

In the London theory, as we will see below (Eq. 3.25), the superconducting current is determined explicitly by the vector potential. Hence this theory is not gauge-invariant. On that ground, F. London concluded that his and his brother theory cannot be generally valid [35].

Ginzburg and Landau claimed that their theory is gauge-invariant. However an expression for the superconducting current in the magnetic field in this theory contains a term identical to Eq. (3.25) [36]. Thus, the GL theory is not gauge-invariant either.

Authors of the BCS theory admit that their theory also is not gauge-invariant, attributing that to an approximate character of their theory [38, 37].

As we saw above, not fulfillment of the gauge invariance is a general property of all magnetizing materials. The fact that the standard theories of superconductivity are not gauge-invariant is a particular case reflecting this property.

1.3 CONDUCTORS

Conductors differ from dielectrics by the presence of conduction or free electrons detached from atoms. Accordingly, as mentioned in Sec. 1.1, these electrons form an additional component of the microscopic current $\mathbf{j}_c$. The latter consists of the total current $\mathbf{j}_t$ and microscopic circular currents $\mathbf{j}'_c$ (see Eq. (1.2)).

The total current in normally conducting media is the transport current; i.e., the current fed by an external power source and existing as long as the power is supplied. This current does not contribute to the magnetization, since the latter is an intrinsic property of the material. Contrarily, $\mathbf{j}'_c$ constitute the circular non-dissipating currents responsible for the Pauli paramagnetism and Landau diamagnetism in metals [39].

Pauli paramagnetism arises from redistribution of the conduction electrons in the momentum space due to interaction of their spin magnetic moment with the field. However, due to Pauli exclusion principle, only electrons with energy close to the Fermi energy can take part in this redistribution. Respectively, contribution of the Pauli paramagnetism in the magnetic susceptibility of a metal is much less than that due to the bound electrons [40].

Landau diamagnetism is associated with the orbital motion of the conduction electrons caused by the field (due to $\mathbf{F}_M$ in Eq.(1.42)). This effect leads to number of remarkable oscillatory phenomena, such as Shubnikov-de Haas and de Haas-van Alphen effects, taking place in very pure metals at low temperatures and high fields [41], and to the quantum Hall effect in 2D specimens [42]. In weak fields magnitude of the magnetic susceptibility due to Landau diamagnetism equals to one-third of that due to Pauli paramagnetism [43].

Hence, the currents associated with $\mathbf{j}'_c$ only slightly vary magnetization in the fields of interest to us. In result, all formulas discussed in this chapter are valid for conducting magnetizing materials.

1.4   PROBLEMS

1.1. Show the equality of magnetic moment of a cylindrical bar magnet with uniform magnetization I to the moment of a solenoid of the same geometry with a current $J = Ic/n$, where n is the number of closely wound turns per unit length. What are the linear current densities in the magnet and in the solenoid? For both objects, the radius R is much smaller than the length L.

Solution. (1) The magnetic moment of the magnet M_m is

$$M_m = I\pi R^2 L.$$

The moment of the solenoid M_s is the sum of the moments of each its turn, which for the given geometry can be considered as a circular current loop, so

$$M_s = \left(\frac{\pi R^2}{c}J\right)nL.$$

At $J = Ic/n$, $M_m = M_s$.

(2) From Eq. (1.22), the linear current density in the magnet is

$$g_m \equiv g_b = cI$$

and that in the solenoid is

$$g_s = nJ$$

Thus, for an external observer the bar magnet (which magnetic moment is due to the moments of microscopic currents in the bar volume) is equivalent to the solenoid (which magnetic moments are due to the moments of the surface current loops) if

$$nJ = cI.$$

1.2. Show that the magnetic part of entropy S_m of a diamagnetic specimen is zero. Use the applied field H_0 parallel to the specimen axis.

Solution. The Maxwell relation in this case (see [16]) reads

$$\left(\frac{\partial S}{\partial H_0}\right)_T = \left(\frac{\partial M}{\partial T}\right)_{H_0}.$$

Since in diamagnets M does not depend on temperature, the specimen entropy S does not depend on the field and it is equal to the entropy at zero field $S_0(T)$. The latter, in its turn is $S_0 = S_{lat} + S_{m0}$, where S_{lat} is the lattice part of the specimen entropy and S_{m0} is its magnetic part associated with disorder in orientation of microscopic (molecular) magnetic moments in zero field. In diamagnetic materials these moments are zero by definition, which implies that $S_m = 0$ regardless on temperature and the field.

Alternatively, the absolute value of S_m can be obtained from statistical mechanics, where entropy of a system is

$$S = k_B \ln \Omega,$$

where k_B is the Boltzmann constant and Ω is number of accessible states of the system. If the system consists of N units, each of which can be with equal probability found in n states, then $\Omega = n^N$.

The magnetic component of the diamagnetic specimen represents a system of identical (both in magnitude and direction) induced magnetic moments of its N molecules. Therefore number of accessible states for the magnetic moment of each molecule is 1 and therefore $\Omega = 1^N = 1$.

Hence,

$$S_m = k_B \ln 1 = 0.$$

1.3. Derive Eq. (1.20).

Solution.. In the presence of total current with the volume density $\mathbf{j}_t$ flowing near the interface shown in Fig. 1.3, Eq. (1.19) takes the form

$$\oint_l H_l dl = (H_{2t} - H_{1t})dL = \int_a (\nabla \times \mathbf{H}) \cdot d\mathbf{a} = \frac{4\pi}{c} \int_a \mathbf{j}_t d\mathbf{a},$$

where we used the Maxwell equation for $\nabla \times \mathbf{H}$ (Ampere's law).

The integral in the right-hand side is the current flowing through the area $dLdz$. If $\mathbf{j}_t$ is finite, as it usually takes place, then in the limit $dz \to 0$ this integral is zero and therefore the condition Eq. (1.18) remains in force. But if the total current flows in a thin layer near the interface, i.e., it represents a surface current of the linear density $\mathbf{g}_t$, then this integral in the limit $dz \to 0$ equals $4\pi g_t/c$, where now g_t is the surface current flowing perpendicular to the page in Fig. 1.3, as it takes place in Eq. (1.21).

As an example, Eq. (1.21) can be used to calculate the field in an infinite solenoid with a current J (the solenoid is in free space). The linear density of this current is nJ (n is the turns' linear density) and field inside the solenoid is

$$H = B = \frac{4\pi}{c} nJ.$$

1.4. An ellipsoidal specimen is in free space with a uniform magnetic field $\mathbf{H}_0$ directed parallel to one of its axes. The magnetic permeability of the specimen material is μ_m and the specimen demagnetizing factor is η.

(a) What are the induction $\mathbf{B}$ and the field intensity $\mathbf{H}$ inside the specimen?

(b) What are magnitudes of the induction B_{ext} and of the field intensity H_{ext} outside the specimen near its surface?

(c) What is the linear density of the surface bound current $\mathbf{g}_b$?

(d) What is the specimen magnetic moment?

Hint: Use Eq. (1.29) and the boundary conditions for H and B.

Solution.

(a) Plugging $B = u_m H$ into Eq. (1.29) we obtain

$$\mathbf{H} = \frac{\mathbf{H}_0}{1 - \eta(1 - \mu_m)}.$$ (p 1.4-1)

and

$$\mathbf{B} = \frac{\mu_m \mathbf{H}_0}{1 - \eta(1 - \mu_m)}.$$ (p 1.4-2)

(b) Denoting the fields on the inner and outer sides of the specimen surface as ($\mathbf{H}$ and $\mathbf{B}$) and ($\mathbf{H}'$ and $\mathbf{B}'$), respectively, for the normal components of $\mathbf{B}$ we write

$$B'_\perp = B_\perp = \frac{\mu_m H_0 \cos\theta}{1 - \eta(1 - \mu_m)}$$

and for the tangential component

$$B'_\parallel = H'_\parallel = H_\parallel = \frac{H_0 \sin\theta}{1 - \eta(1 - \mu_m)}.$$

Hence, the magnitude of the external field near the specimen surface is

$$B_{ext} = H_{ext} = \frac{H_0}{1 - \eta(1 - \mu_m)}\sqrt{\mu_m^2 \cos^2\theta + \sin^2\theta}$$ (p 1.4-3)

And the angle φ between H_{ext} and the normal to the surface is

$$\tan\varphi = \frac{B'_\parallel}{B'_\perp} = \frac{\tan\theta}{\mu_m}$$ (p 1.4-4)

(c) Using Eq. (1.22) we obtain

$$\mathbf{g}_b = -\frac{c}{4\pi}\frac{(1 - \mu_m)\sin\theta}{1 - \eta(1 - \mu_m)}\mathbf{H}_0 \times \mathbf{n}.$$ (p 1.4-5)

(d) M calculated from definition (1.3) is

$$\mathbf{M} = -\frac{V}{4\pi}\frac{(1 - \mu_m)}{1 - \eta(1 - \mu_m)}\mathbf{H}_0$$ (p 1.4-6)

For superconductors in the Meissner state ($\mu_m = 0$), Eqs. (p1.4-1)–(p1.4-6) yield

$$\mathbf{H} = \frac{\mathbf{H}_0}{(1 - \eta)}$$

$$\mathbf{B} = 0$$

$$B_{ext} = H_{ext} = \frac{H_0 \sin\theta}{(1 - \eta)}$$

$$\varphi = \pi/2$$

The external field is tangential to the specimen surface, in other words it bends around the specimen.

$$\mathbf{g}_b = -\frac{c\sin\theta}{4\pi(1-\eta)}\mathbf{H}_0 \times \mathbf{n}$$

and

$$\mathbf{M} = -\frac{V\mathbf{H}_0}{4\pi(1-\eta)}$$

After all, for specimens of cylindrical geometry these quantities are

$$\mathbf{H} = \mathbf{H}_0,$$

$$\mathbf{B} = 0,$$

$$B_{ext} = H_{ext} = H_0,$$

$$\varphi = \pi/2,$$

$$\mathbf{g}_b = -\frac{c}{4\pi}\mathbf{H}_0 \times \mathbf{n},$$

$$\mathbf{M} = -\frac{V\mathbf{H}_0}{4\pi}.$$

1.5. Coming from the Biot–Savart law $(d\mathbf{H} = J(d\mathbf{l} \times \mathbf{R})/cR^3$, where $\mathbf{R}$ is the radius-vector from the current element $Jd\mathbf{l}$ to a point of observation [18, 19, 22]), show that the intensity of a magnetic field away from a system of currents with density $\mathbf{j}$ confined in a volume V, if the permeability of the space around V is unity, is equal to

$$\mathbf{H} = \mathbf{B} = \nabla \times \mathbf{A},$$

where $\mathbf{A}(= \mathbf{A}_B)$, the vector potential, has the form given in Eq. (1.35).

Hint: The current element $Jd\mathbf{l} = \mathbf{j}dV$; use a vector identity $\nabla \times (f\mathbf{F}) = f(\nabla \times \mathbf{F}) + (\nabla f) \times \mathbf{F}$ and take into account that $\nabla(1/R) = -\mathbf{R}/R^3$.

1.6. Derive the formula for $\mathbf{B}$ given in Eq. (1.37) using the vector potential from Eq. (1.36).

Hint: Use the "back of the cab" vector identity $\mathbf{A} \times (\mathbf{B} \times \mathbf{C}) = \mathbf{B}(\mathbf{A}\cdot\mathbf{C}) - \mathbf{C}(\mathbf{A}\cdot\mathbf{B})$.

1.7. Derive $\mathbf{B}$ using the scalar potential Eq. (1.38).
Hint: $\nabla(1/R) = -\mathbf{R}/R^3$.

1.8. Derive an expression for the magnetic moment of a wire loop carrying a current J; the loop is flat and the current is linear.

Solution. In this case, the current element $\mathbf{j}_{mic}dV = Jd\mathbf{l}$ and, accordingly, Eq. (1.1) takes the form

$$\mathbf{M} = \frac{1}{2c}\oint_l \mathbf{r} \times d\mathbf{l} = \frac{J\mathbf{a}}{c},$$

where $|\mathbf{a}|$ is the area bound by the loop, and the direction of $\mathbf{a}$ is determined by the right-screw rule.

1.9. Show that the vector potentials Eqs. (1.43), (1.46) and (1.47) differ from each other by a gradient of a function of coordinates. What are these functions?

Answer. $\mathbf{A}_2 = \mathbf{A}_1 - \nabla f$, $\mathbf{A}_3 = \mathbf{A}_1 + \nabla f$, $\mathbf{A}_2 = \mathbf{A}_3 - 2\nabla f$, where $f = Hxy/2$, the field $\mathbf{H}$ is parallel to $z-$axis, and $\mathbf{A}_1$, $\mathbf{A}_2$, $\mathbf{A}_3$ are the vector potentials in Eqs. (1.43), (1.46), (1.47), respectively.

1.10. A magnetic field $\mathbf{H}_0$ is parallel to one of axes of a superconducting specimen. The specimen is in the Meissner state ($\mathbf{B} = 0$) and its demagnetizing factor is η. Using Eq. (1.29) find (a) the field intensity $\mathbf{H}$ inside the specimen; (b) the specimen magnetic moment $\mathbf{M}$; (c) the outer field $\mathbf{H}_{ext}$ near the specimen surface.

Solution.

(a). Since $\mathbf{H}_0$ is parallel to the specimen axis, $\mathbf{H}$ is parallel to $\mathbf{H}_0$. Then, from Eq. (1.29) the magnitude of $\mathbf{H}$ is

$$\mathbf{H} = \frac{\mathbf{H}_0}{(1 - \eta)},$$

(b).

$$\mathbf{M} = IV = V\frac{(\mathbf{B} - \mathbf{H})}{4\pi} = -V\frac{\mathbf{H}_0}{4\pi(1 - \eta)}.$$

(c). Due to the boundary condition for B_n Eq. (1.17) and the Meissner condition, $B = 0$, the outer field near the specimen H_{ext} is tangential to its surface, i.e., the field lines bend around the specimen (see Fig. 3.10). Tangential component of the inner field $\mathbf{H}$ is $H\sin\theta$, where θ is the angle between the normal to the surface and $\mathbf{H}_0$. Hence, from the boundary condition for H_t Eq. (1.18), we find

$$H_{ext} = H\sin\theta = \frac{H_0}{(1 - \eta)}\sin\theta.$$

1.11. (a) Using formula for the field of a magnetic dipole Eq. (1.37) and the boundary condition for B_n Eq. (1.17), calculate magnetic moment $\mathbf{M}$ of a spherical specimen in the Meissner state. (b) Calculate the moment of the same sphere using the boundary condition for H_t Eq. (1.18). (c) Show validity of Eq. (1.29). (d) Find the magnetic field outside the specimen. Radius of the specimen is R_0 and demagnetizing factor $\eta = 1/3$.

Solution

(a) Outside the specimen the field $\mathbf{H}'$ is

$$\mathbf{H}' = \mathbf{B}' = \mathbf{H}_0 + \mathbf{H}_M.$$

From the boundary condition Eq. (1.17) along with the Meissner condition $\mathbf{B} = 0$ inside the specimen it follows that the normal component of B' at $R = R_0$ is zero. Hence, we write

$$H_0 \cos\theta + \frac{3MR_0 \cos\theta}{R_0^5} R_0 - \frac{M \cos\theta}{R_0^3} = 0$$

So,

$$M = -\frac{R_0^3}{2} H_0 = -\frac{3}{2} \frac{V}{4\pi} H_0 = -\frac{V}{4\pi(1 - 1/3)} H_0,$$

where the negative sign means that $\mathbf{M}$ is anti-parallel to $\mathbf{H_0}$.

On the other hand, by definition, M of our specimen at the Meissner state is

$$M = VI = V\frac{(B - H)}{4\pi} = -V\frac{H}{4\pi}.$$

Comparing the last two expressions, we see that

$$H_0 = (1 - 1/3)H = (1 - \eta)H.$$

This is identical to Eq. (1.29) for $B = 0$.

(b). For the tangential component of H_t (with $H = 3\mathbf{H_0}/2$) outside and inside this sphere we write:

$$H_0 \sin\theta + 0 - \frac{M \sin\theta}{R_0^3} = \frac{3}{2} H_0 \sin\theta$$

So,

$$M = -\frac{R_0^3}{2} H_0,$$

which is the same as that obtained above.

(b) The field outside the specimen is

$$\mathbf{B}' = \mathbf{H}' = \mathbf{H_0}(1 + \frac{R_0^3}{2r^3}) - \frac{3R_0^3}{2r^5}(\mathbf{H_0} \cdot \mathbf{r})\mathbf{r}$$

where $r \geq R_0$.

We leave it for the reader to show that $\mathbf{B}'(R_0)$ is the same as $\mathbf{H}_{ext}$ calculated in problem 1.10.

CHAPTER 2

GYROMAGNETIC RATIO AND RELATED PHENOMENA

Each closed microscopic current in the media (like any other closed current) in addition to the magnetic moment μ possesses an angular momentum ι (iota) proportional to it. The momentum ι is either parallel or antiparallel to μ depending on the sign of the current carriers. Due to this fact the closed microscopic currents are also microscopic gyroscopes; in particular, being in a magnetic field with intensity $\mathbf{H}$, they experience a torque $\tau = \mu \times \mathbf{H}$ and, therefore, precess. The precession angular frequency Ω is equal to τ/ι [44].

If the precession is slow, it is referred to as the Larmor precession and its frequency as the Larmor frequency o. The slowness condition, called Larmor's condition, which will be specified below, implies that the Larmor frequency of orbiting atomic electrons must be much less than 10^{16} Hz or $H \ll 10^9$ G. In all stationary fields that can be achieved in laboratories, this condition is well satisfied. It is also well satisfied for spin magnetic momenta of electrons and nuclei, and for the orbital momenta of electrons in Cooper pairs of superconductors. Therefore, it is safe to say that in a magnetic field the closed microscopic currents in any medium precess with the Larmor frequency. In the vector form[1] the latter is expressed as

$$o = -\gamma\mathbf{H}, \tag{2.1}$$

where γ is a gyromagnetic ratio equal to μ/ι and the negative sign reflects the fact that the induced magnetic moment caused by the precession is directed against $\mathbf{H}$.

---

[1] The direction of the vector of angular frequency (or angular velocity) is determined by the right screw rule relative to the circular current.

DOI: 10.1201/9781003355786-2

In general case

$$\gamma = \frac{eg}{2mc} \tag{2.2}$$

where g is the g- or Landé factor.

The g-factor of orbiting electrons with compensated spins, the precession of which determines magnetization of diamagnetics, is equal to unity. This fact underlies the Kikoin-Goobar effect in superconductors [45]. The Landé factor of free atoms with both orbital and spin degrees of freedom is greater than 1 and fractional; it determines the fine structure of Zeeman spectra [39]. The g-factor of the electron spin is equal (very close) to 2. The gyromagnetic ratio of uncompensated electron spins is responsible for the Barnett effect. Precession of the orbiting electrons with uncompensated spin leads to the Einstein-de Haas effects and lies in the foundation of electron paramagnetic/spin resonance (EPR or ESR) [46]. Nuclei and elementary particles with non-zero spin, such as proton, neutron and muon, also precess in magnetic field and correspondingly posses the gyromagnetic ratio. Precession of these particles provides NMR (Nuclear Magnetic Resonance) and μSR (Muon Spin Resonance/Rotation/Relaxation) [47].

All magnetic properties of media in one or another way owe their existence to gyromagnetic ratio of closed microscopic currents inside them. In this chapter we discuss experimental and theoretical aspects of phenomena associated with the gyromagnetic ratio of electrons.

2.1 LARMOR THEOREM

The Larmor theorem [48] takes a key part in understanding of magnetism both in normal and superconducting materials.

Consider Bohr's model of the atom, a single spinless electron orbiting a massive nucleus under the action of the central force $\mathbf{F}_c$ at zero field. The nucleus is in an origin of an inertial coordinate system[2] S; the orbit radius is R; the electron velocity is $\mathbf{v} = \boldsymbol{\omega} \times \mathbf{R}$, where $\boldsymbol{\omega}$ is the angular velocity of the electron; and the orbital magnetic moment $\boldsymbol{\mu}$ is

$$\boldsymbol{\mu} = \frac{1}{c}\pi R^2 \left(\frac{e}{2\pi}\boldsymbol{\omega}\right) = \frac{e}{2mc}(mR^2\boldsymbol{\omega}) = \frac{e}{2mc}\boldsymbol{\iota} \equiv \gamma\boldsymbol{\iota}. \tag{2.3}$$

We see that g-factor of the orbiting spinless electron is unity and $\boldsymbol{\mu}$ is directed against $\boldsymbol{\iota}$ since $e < 0$.

The magnitude of the electron linear velocity in the Bohr model $v = \alpha c$, where $\alpha (= e^2/\hbar c$ with $\hbar$ denoting the reduced Planck constant) is the fine-structure constant equal to $1/137$. Hence, relativistic effects are negligible and therefore the equation of motion of this electron is

$$m\frac{d\mathbf{v}}{dt} = \mathbf{F}_c. \tag{2.4}$$

---

[2]More correctly, of the coordinate system belonging to the inertial reference frame.

In a static magnetic field of intensity $\mathbf{H}$ (assumed parallel to the z axis), the electron experiences the action of the force $\mathbf{F}_M$ (Eq. (1.41)) in addition to $\mathbf{F}_c$. The equation of motion in S becomes

$$m\frac{d\mathbf{v}}{dt} = \mathbf{F}_c + \frac{e}{c}\mathbf{v} \times \mathbf{H}. \tag{2.5}$$

For an electron orbiting in a plane transverse to $\mathbf{H}$, an action of the field is well known: depending on the direction of orbiting, the electric field induced by the changing magnetic field either increases or decreases the electron velocity, in both cases leading to an appearance of the field induced moment $\boldsymbol{\mu}_i$ directed against $\mathbf{H}$; herewith the radius of the electron orbit remains unchanged [22, 49]. To find out the action of $\mathbf{F}_M$ on all bound electrons consider their motion in a coordinate system S' in which z' axis coincides with the z of S. Now let S' rotate about this axis with an angular frequency $\Omega \ll \omega$.

The system S' is not inertial, so the equation of motion of the electron in this system is [44]

$$m\frac{d\mathbf{v}'}{dt} = \mathbf{F}_c + \frac{e}{c}\mathbf{v}' \times \mathbf{H} + 2m\mathbf{v}' \times \boldsymbol{\Omega} + m\boldsymbol{\Omega} \times (\mathbf{r} \times \boldsymbol{\Omega}), \tag{2.6}$$

where $\mathbf{v}'$ is the electron velocity in S' and the last two terms are, respectively, Coriolis and centrifugal pseudo forces caused by the non-inertia of the S'.

The condition $\Omega \ll \omega$ implies that

$$\mathbf{v}' = \mathbf{v} - \boldsymbol{\Omega} \times \mathbf{R} = \boldsymbol{\omega} \times \mathbf{R} - \boldsymbol{\Omega} \times \mathbf{R} \approx \mathbf{v}. \tag{2.7}$$

So, in this approximation the centrifugal force in Eq. (2.6) is negligible (since it is much less than the Coriolis force) and therefore the orbital radius R remains unaffected (see also [18, 22, 50]). Consequently, the magnitude of the orbital magnetic moment μ is unaffected either. In turn, Eq. (2.6) becomes

$$m\frac{d\mathbf{v}}{dt} = \mathbf{F}_c + \frac{e}{c}\mathbf{v} \times \mathbf{H} + 2m\mathbf{v} \times \boldsymbol{\Omega}. \tag{2.8}$$

Replacing $\boldsymbol{\Omega}$ by $\boldsymbol{o}$ with $g = 1$ (Eqs. (2.1) and (2.2)), we see that the equation of motion of the electrons (regardless of the plane of its circulation) in the rotating coordinate system S' is the same as it was in the stationary S system at zero field (Eq. (2.4)). This means that (a) rotation with frequency $\boldsymbol{o}$ is equivalent to the appearance of a field $\mathbf{H}' = -(2mc/e)\boldsymbol{o}$ in the S', which annihilates the field in this system; and (b) in the S system the entire construction (atom) rotates with the Larmor frequency. The latter implies the appearance of a field-induced circular current of the Larmor frequency, which magnetic moment is directed opposite to $\mathbf{H}$.

On the other hand, if $\mathbf{H} = 0$, then the rotation with a frequency $\boldsymbol{\omega}_r$ (any frequency much less than the orbital frequency ω) generates the field $\mathbf{H}_B = -(2mc/e)\boldsymbol{\omega}_r$ directed parallel to $\boldsymbol{\omega}_r$. Existence of this field was for the first time demonstrated by Barnett, so we will call it the Barnett field.

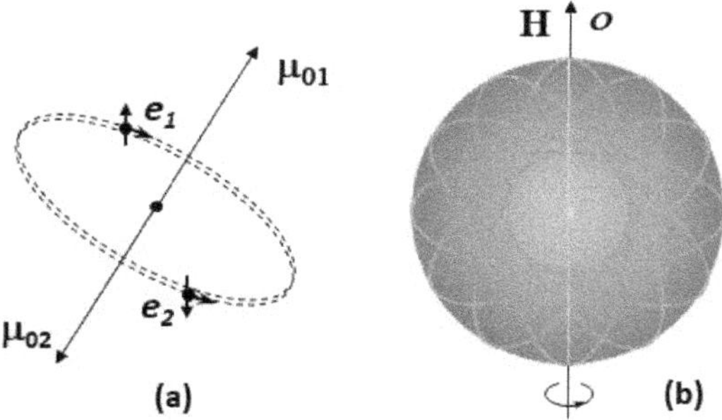

Figure 2.1 (a) A semi-classical model of a helium atom in zero field. Dashed circles are trajectories of the orbiting electrons; small vertical and tangential arrows represent electron spins and velocities, respectively; μ_{01} and μ_{02} are magnetic moments due to the orbital motion of the 1st and 2nd electron, respectively. (b) time- and space-averaged image of a diamagnetic atom in the magnetic field $\mathbf{H}$; o is the angular velocity of the atom caused by the Larmor precession of the electron orbits.

Important to stress that the rotation-induced Barnett field is a *real magnetic field* of the intensity $\mathbf{H}_B$ (Barnett calls it "intrinsic magnetic intensity of rotation" [55]), which produces magnetization due to the work done by an external agent. For example, if a demagnetized specimen inserted into the pickup coil of a magnetometer is set in rotation, it will acquire a magnetic moment which will be registered by the magnetometer. This phenomenon, known as the Barnett effect, is discussed in the next section. In Sec. 2.3.3 we will see that the field $\mathbf{H}'$ (the Barnett field with $\omega_r = o$) is also responsible for paramagnetism. In superconductors, this field causes a phenomenon called the London moment, which we will discuss in Ch. 5.

Overall, the Larmor precession represents the common rotation of a system of circulating identical charges[3] about the axis parallel to $\mathbf{H}$, which is superimposed on the motion of these charges at zero field. This statement constitutes the essence of the Larmor theorem; the condition for its fulfillment is $|o| \ll |\omega|$ or $|\mathbf{v}' - \mathbf{v}| \ll |\mathbf{v}|$. However, this is not the end of the story.

Fig. 2.1a shows a semi-classical model of the helium atom in zero field. We took He because it is a simple example of diamagnetic atoms. The time-mean motion of electrons represents circular currents centered at the nucleus with corresponding magnetic moments μ_0. As it takes place in all diamagnetic atoms, the spin and orbital magnetic moments are compensated. Hence, electrons are effectively spinless, while the orbital momenta μ_{01} and μ_{02} are directed opposite. At the same time the line $\mu_{01} - \mu_{02}$ in Fig. 2.1a can be oriented in any way due to 3D symmetry of the medium in absence of the magnetic field. This means that the currents averaged over atoms containing in, e.g., a unit volume of a specimen represent a spherical perfectly conducting

---

[3]More precisely, of the charges with identical charge-to-mass ratio.

shell formed by the dissipation-free currents with mutually compensated magnetic moments[4]. Then the Larmor theorem states that in the field the entire shell rotates with the angular frequency o as shown in Fig. 2.1b. This results in the field-induced circular current, which magnetic moment is negative; i.e., it directs against **H**.

So, for a statistical ensemble of the spinless atoms[5] in thermodynamic equilibrium the Larmor theorem can be formulated as follows: the net effect of a static magnetic field is the appearance of field-induced circular currents of Larmor's frequency in each atom; all the induced currents lie in planes transverse to **H** and their magnetic moments are negative.

Thus, the Larmor theorem helps to explain the phenomenon of diamagnetism. To remind, the discovery of this phenomenon by Faraday, who coined the term diamagnetism [51][6], was a major defeat of the Poisson theory.

On the other hand, we understand that the picture in Fig. 2.1b looks as a miracle for an observer in the S system. Imagine a student being in this system. In absence of the field, she sees a motionless spherical shell. Then she turns the field on and the shell immediately starts rotating with a constant speed. Increasing the field makes this speed faster, while decreasing it - slower. Herewith, the rotational speed is entirely determined by the field magnitude regardless on the rate and sign of its change. Reversing the field polarity momentary reverses direction of rotation, and turning the field off instantly stops it. This picture can be even more impressive if we imagine each atom as a spinning top. At zero field (if we neglect the thermal motion) our student is surrounded by myriads of such tops with motionless and arbitrary oriented axes. Suddenly, something happens (someone turns on the field): at once, all the tops begin synchronously precessing relative to invisible axes passing through the center of mass of each top and parallel to the field. Increasing the field speeds the precession up, and so on.

Definitely, our student is totally flabbergasted because the observing picture denies (or looks as denying) the most fundamental physics laws: the Galilean law of inertia (the angular speed changes discontinuously); the Faraday law of induction (an induced current should depend on the rate and sign of the field *change*, but it does not); and the law of energy conservation, since the static magnetic field cannot do work!

Our student decides to sort out with the Larmor theorem. Let us help her.

---

[4]Such a model of a diamagnetic atom was proposed by Maxwell, who proceeded from the Ampère hypothesis [1].

[5]This can be molecules, ions or any other microscopic units constituting the ensemble.

[6]At the end of this article Faraday writes: "Feb. 2, 1846. I add the following notes and references to these Researches: - Brugmans first observed the repulsion of bismuth by a magnet in 1778. *Antonii Brugmans Magnetismus de affinitatibus magneticis observationes magneticœ*. Lugb. Betav. 1778. M. Le Baillif on the Repulsion of a Magnet by Bismuth and Antimony, Bulletin Universel, 1827, vol. vii, p. 371; vol. viii. pp. 87. 91. 94; Saigey on the Magnetism of certain natural combinations of Iron, and on the mutual repulsions of Bodies in general. Ibid. 1828, vol. ix. pp. 89. 167. 239; Seebeck on the Magnetic Polarity of different Metals, Alloys and Oxides. Ibid. 1828, vol. ix. p. 175".

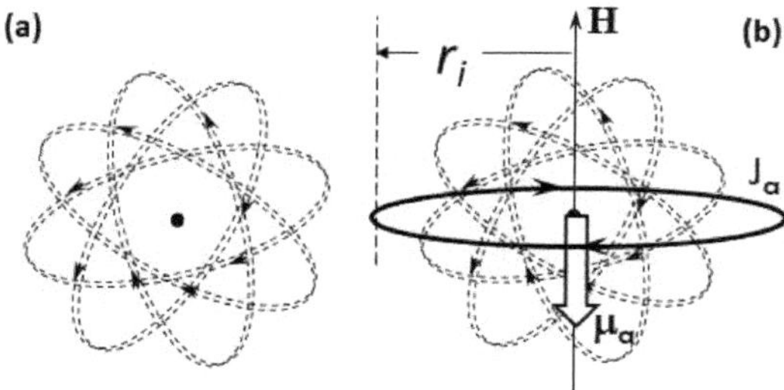

Figure 2.2 A semi-classical image of the valence shell of the Ne atom at zero (a) and non-zero field **H** (b). Dashed circumferences depict currents due to individual electrons at zero field; arrows indicate directions of the currents. In (b), the solid circumference in the plane transverse **H** depicts the field-induced current in the atom $J_a = ZJ_i$, where Z is the total number of electrons in the atom (atomic number) and J_i is the induced current per one electron; r_i is the root mean square radius of the induced currents (see Sec. 1.2.7); the fat arrow down is the induced magnetic moment of the atom $\boldsymbol{\mu}_a$.

For that we consider again a non-conducting specimen consisting of atoms with filled electron shells. Such a specimen is diamagnetic, all electrons are bound to atoms with even number of electrons. The spin magnetic moments of electrons in each atom are mutually compensated, so the electrons are effectively spinless. In the field absence, the orbital magnetic moments in each atom are mutually compensated either, so M_a, the magnetic moment of an atom, is

$$\mathbf{M}_a = \sum_a \boldsymbol{\mu}_0 = 0 \tag{2.9}$$

where $\boldsymbol{\mu}_0$ is an orbital magnetic moment of a single electron in zero field and the summation is taken over all electrons of the atom. A semi-classical image of a diamagnetic atom at zero field is shown in Fig. 2.2a.

Since our atom does not move[7], a sum of instantaneous velocities of electrons at zero field is also zero[8], i.e.,

$$\sum_a \mathbf{v}_0 = 0. \tag{2.10}$$

When $\mathbf{H}_0$ is turned on, the field intensity in the specimen changes from zero to **H** and each electron is subject to the Lorentz force in addition to the forces acting on it at zero field. In our case (see Eq. (1.41)), the Lorentz force

---

[7]The atom can move due to temperature. However Eq. (2.9) remains in force since the atom moves as a whole.

[8]This also follows from 3D symmetry of medium at zero field.

is

$$\mathbf{F}_L = -\frac{e}{c}\left(\frac{\partial \mathbf{A}}{\partial t}\right) + \frac{e}{c}\mathbf{v} \times \mathbf{H} =$$

$$-\frac{e}{c}\left(\frac{\partial \mathbf{A}}{\partial t}\right) + \frac{e}{c}\mathbf{v} \times (\nabla \times \mathbf{A}), \quad (2.11)$$

where $\mathbf{A}$ is the vector potential of the $\mathbf{H}$-field (Eq. (1.40)).

The second term in this formula is the magnetic force $\mathbf{F}_M$ which is transverse to velocity. Therefore, $\mathbf{F}_M$ does not change the velocity magnitude but provides a torque $\boldsymbol{\tau} = \boldsymbol{\mu}_0 \times \mathbf{H}$. Consequently, due to non-zero ι_0, the magnetic moment of each electron in the atom (and therefore in the specimen) precesses about the direction of $\mathbf{H}$. Precession starts simultaneously all over the specimen with the Larmor frequency, which does not depend on parameters of the electron orbits, and themselves parameters (the magnitudes of the orbital radius R_0, speed v_0 and, therefore, of the moment μ_0) are unaffected by the field, provided the Larmor condition is met. Hence, all orbital momenta in each atom (and in the specimen as a whole) precess synchronously, retaining the mutual orientation that they had in zero field. Therefore Eqs. (2.9) and (2.10) hold regardless of the field.

Thus, the Larmor precession (the motion occurring due the force $\mathbf{F}_M$) *does not* change the magnetic moment of the atoms and therefore the moment of the specimen: all of them remain to be zero, as it must be the case because the magnetic force does no work. As is well known from the Faraday law, the work of a magnetic field (not to be confused with the magnetic force) occurs due to the action of an electric force $\mathbf{F}_E$ arising from an electric field $\mathbf{E}_i$ induced by a changing magnetic field [20]. This force, determined by the first term of the Lorentz force $\mathbf{F}_L$, is equal to

$$\mathbf{F}_E = e\mathbf{E}_i = -\frac{e}{c}\left(\frac{\partial \mathbf{A}}{\partial t}\right). \quad (2.12)$$

The fact that the atomic structure is unaffected by $\mathbf{F}_M$ means that $\mathbf{F}_E$ is the net (resultant) force acting on the electron. So, Newton's acceleration law for this electron is

$$\frac{\partial \mathbf{p}}{\partial t} = m\frac{\partial \mathbf{v}}{\partial t} = -\frac{e}{c}\left(\frac{\partial \mathbf{A}}{\partial t}\right), \quad (2.13)$$

where $\mathbf{p}$ is the kinetic linear momentum, and the partial derivative in Newton's law is used because the momentum $\mathbf{p}$, like the vector potential $\mathbf{A}$, depends on time and on coordinates.

However, the time in Eq. (2.13) cancels out, resulting in a time-independent relationship between changes in $\mathbf{A}$ and $\mathbf{v}$, namely

$$d\mathbf{v} = -\frac{e}{cm}d\mathbf{A}. \quad (2.14)$$

When the magnet is turned on, the vector potential changes from zero[9] to $\mathbf{A}$, corresponding to the change of the intensity from zero to $\mathbf{H}$. Herewith the force $\mathbf{F}_E$ changes the electron velocity from $\mathbf{v}_0$ to $\mathbf{v}'$. The difference (assumed small) $\mathbf{v}_i = \mathbf{v}' - \mathbf{v}_0$ is the velocity induced due to the applied field. From Eq. (2.14) this velocity is

$$\mathbf{v}_i = -\frac{e}{cm}\mathbf{A}. \tag{2.15}$$

The formula was discussed in Sec. 1.2.7, where it was shown that the circular gauge (Eq. 1.43) is the only appropriate gauge for the vector potential in media. Having that, Eq. (2.15) is rewritten as

$$\mathbf{v}_i = -\frac{e}{2cm}\mathbf{H} \times \mathbf{r}, \tag{2.16}$$

where $\mathbf{r}$ is the radius of the induced circular motion of an electron lying in the plane transverse to $\mathbf{H}$ and centered in the nucleus.

Accordingly, the angular velocity of the induced circular motion $\boldsymbol{\omega}_i (= \mathbf{r} \times \mathbf{v}_i)/r^2)$ is

$$\boldsymbol{\omega}_i = -\frac{e}{2mc}\mathbf{H}. \tag{2.17}$$

Comparing $\boldsymbol{\omega}_i$ with $\boldsymbol{o}$ in Eq. (2.1) we see that it equals to the Larmor frequency with $g = 1$. Hence, as was already pointed out in Sec. 1.2.7, we really deal with precession of the orbiting spinless electron, because in *all* other cases $g \neq 1$.

Accordingly, the induced current J_i in the orbit of a single electron is

$$J_i = \frac{e}{2\pi}\omega_i = \frac{e^2}{4\pi mc}H. \tag{2.18}$$

Both $\boldsymbol{\omega}_i$ and J_i are independent on r, the radius of the field-induced motion proportional to R_0. However, it is not the case for the induced magnetic moment μ_i and the change of the electron kinetic energy e_i, they both depend on r^2. However, r in the orbits of different orientation (relative to $\mathbf{H}$) are different. So, to calculate the induced magnetic moment in an atom we have to use the mean square $\langle r^2 \rangle$, which we will denote as r_i^2. Then an average induced magnetic moment per electron written as a function of H is[10]

$$\mu_i = \frac{\pi r_i^2}{c}\frac{e}{2\pi}\omega_i = -\frac{e^2 r_i^2}{4mc^2}\mathbf{H}. \tag{2.19}$$

So, the induced magnetic moment in an electron orbit is always (regardless of either the orientation of μ_0 or the sign of e) directed against $\mathbf{H}$, as it must be the case in diamagnetics. A semi-classical image of a polarized diamagnetic atom is shown in Fig. 2.2b. Overall, the net effect of a magnetic field on a diamagnetic specimen is the appearance of the induced circular currents in

---

[9]When the magnet is off, $\mathbf{A}$ is taken to be zero.

[10]μ_i in terms of $\mathbf{A}$ is given in Eq. (1.48).

each atom lying in the transverse to $\mathbf{H}$ planes. The frequency of the induced currents is equal to the Larmor frequency and their magnetic momenta are negative. At the same time, since r is proportional to R_0 and the latter does not depend on the field, $r_i (= \sqrt{\langle r^2 \rangle})$ does not depend on the field either, provided $v_i \ll v_0$. The same result follows from the traditional derivation of Larmor's theorem given above.

Thus, we have proved the Larmor theorem without switching the reference frames. Diamagnetism is inherent in all media, but in paramagnetics it is outshined by the magnetization caused by non-zero intrinsic atomic magnetic moments. As mentioned, paramagnetism arises from the tendency of atomic moments to align along the field because such an alignment decreases free energy of the system (specimen). As shown in Sec. 2.3.2, this tendency is also the result of the precession of electron orbits following from the Larmor theorem.

Finally, our student can relax, as she has become convinced that her so strange observations are fully consistent with the laws of classical physics.

Indeed, there are no conflicts both with the energy conservation and the Faraday law. The induced current and respectively magnetization in diamagnetics is a consequence of the Faraday law of induction (the first part in Eq. (2.11)), as it was anticipated by Weber [52] and Maxwell [1]. The energy to magnetize the specimen is provided by the work done by the vortex electric field induced by the changing magnetic field[11].

Another mystery concerns the apparent non-inertia of the Larmor precession. The point here is that as seen from Eq. (2.14), the time drops out at its integration implying that the induced velocity (Eqs. (2.15)) does not depend on the rate of the field change. This is the feature that can be interpreted as the inertia-free motion. It should not be forgotten that the derived formulas are valid under the condition $\omega_i \ll \omega_0$ or the period of the induced rotation T_i is much greater than T_0, the period of the electron orbiting. Taking $r_0 \sim 10^{-8}$ cm, we find that $T_0 (= 2\pi r_0 / c\alpha)$ is of the order of 10^{-16} s. The period T_0 represents a time-scale of establishing equilibrium in the atomic system. Hence, any real (observable) time interval of the field change exceeds T_0 by lots of orders of magnitude. In other words, the real field change is always a slow adiabatic (i.e., reversible) process. Respectively, at any instant of observation our student registers the equilibrium value of the angular velocity, in full consistency with the law of inertia and thermodynamics.

On the other hand, the absence of time in Eq. (2.15) indicates that all microscopic currents in the Universe are constantly precessing. Herewith, the speed $\mathbf{v}_i$ of a charge forming any of these current is controlled by the vector potential of the field acting on it, which is the field $\mathbf{H}$[12].

---

[11]Ultimately, this energy comes from the magnet power supply, which does the work required to change the applied magnetic field.

[12]From Eq. (2.15) it follows that a registered today $\mathbf{v}_i$ could have been achieved over the entire lifetime of the Universe.

Before completing this section let us summarize the differences between the first (traditional) derivation of the Larmor theorem (with different reference frames) and the second one, based on the Lorentz force. (1) The traditional derivation does not explain where the energy of the induced rotation comes from; the second derivation does this. (2) From the traditional derivation it seems obvious (or intuitively obvious) that the rms radius of the induced current is the same as the rms radius of the atom R_a (roughly, the radius of the electron cloud surrounding the nucleus). In the second derivation this is not necessarily the case: r_i and R_a, though related, are different quantities and therefore can be not equal. Below we will see, that experimental data support this conclusion.

2.2   GYROMAGNETIC EFFECTS

The gyromagnetic effect is a macroscopic phenomenon caused by the coupling between the magnetic moment and the angular momentum of microscopic units of matter. The effect directly demonstrates this coupling and allows one to measure the gyromagnetic ratio γ. The gyromagnetic effect in ferromagnetics was independently discovered by Barnett in 1914 [53] and Einstein and de Haas in 1915 [54]. These groundbreaking studies, along with a significant number of experiments that followed (see [55]), played a decisive role in establishing the electrodynamic nature of magnetization, determining the intrinsic electron magnetic moment, and paved the way to discover the magnetic resonance phenomena. Of similar importance for diamagnetic materials is the discovery of the gyromagnetic effect in superconductors by I. Kikoin and Goobar in 1938 [45].

An idea of an existence of the gyromagnetic effect is based on celebrated hypotheses of Ampère about non-dissipating molecular electric currents [56] and Weber about the inertia of electric charges [52]. Coming from that, the magnetic elements of a ferro- and paramagnetic specimen, should behave simultaneously as magnets and gyroscopes.

A gyroscopic model of an atom with non-zero angular momentum is shown in Fig. 2.3. If the wheel W spins rapidly about the axis AA, and the entire frame rotates slowly (so that the action of the centrifugal force can be neglected) around the vertical axis C, then the wheel moves up or down so that the direction of its angular velocity approaches the direction of the applied rotation around the C axis. In the ideal case (in the absence of interactions with neighbors, conditionally denoted by the springs S) the angle θ becomes either zero or π depending on direction of the wheel spinning. Hence, the alignment of the spinning axis with the axis of the applied rotation decreases kinetic energy of the wheel. This means that, in contrast to diamagnetics, magnetization of a para- and ferromagnetics *decreases* their free energy. In other words, magnetization of these materials represents a magnetocaloric effect or

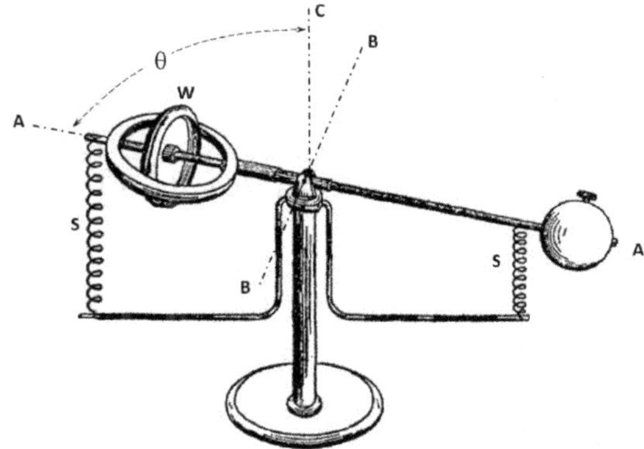

Figure 2.3 A balanced gyrostat as a model of an atom with non-zero angular momentum. W is a wheel rapidly spinning about the axis AA; without the springs SS, the axis AA is free to move in altitude with respect to the horizontal axis BB; θ is the angle which AA makes with the vertical axis C; the springs SS represent resistive forces occurring due to interaction with other atoms; the axis BB along with the entire framework can rotate with respect to the axis C. After Barnett [55]; reprinted with permission from the American Physical Society.

reversed adiabatic demagnetization, which is broadly used in low-temperature studies (see, e.g., problems after Ch. 2 in [16]).

Barnett [55] distinguishes four gyromagnetic effects. Auwers [57] extends this list to eight positions. Below we dwell on the main historical facts of the measurements of the gyromagnetic effects and discuss in more detail the experiment of I. Kikoin and Goobar.

2.2.1 The gross magnet as a gyrostat

In a strongly magnetized bar magnet all magnetic elements are parallel to the bar axis. If the gyroscopic atomic model is correct, each of these elements has an angular momentum aligned along the same axis. Therefore, the whole bar possesses a concealed angular momentum, i.e., it represents a gyrostat[13]. Then, if the bar magnet is pivoted like the rod AA in Fig. 2.3 and set in rotation about the C axis, the angle θ should either decrease or increase depending on the sign of γ. A very sensitive apparatus based on this principal was designed and constructed by Maxwell in 1861[14]. However, the experiment failed due to disturbing actions of the earth's magnetic field (the chief difficulty) and other extraneous impacts masking the effect is sought for [1].

---

[13]The gyrostat is a gyroscope with concealed angular momentum.

[14]Kapitza recollects that he found this apparatus in one of the cabinets of the Cavendish Laboratory in 1920s [58].

2.2.2 Magnetization by rotation (Barnett effect)

The effect of magnetization by rotation, following from the above idea of Maxwell, was discovered by Barnett [53, 55]. If a demagnetized bar magnet is brought into rotation about its axis, then, according to the gyroscopic atomic model, all chaotically oriented magnetic elements should strive to orient themselves along the axis of the applied rotation; in result the bar should be magnetized. Herewith, at sufficiently slow rotation, the rotation-induced magnetization should be proportional to the rotation speed. Hence, by measuring this magnetization and knowing the rotation speed, one can determine the sign and magnitude of the gyromagnetic ratio. As Barnett showed [55], magnetization by rotation is equivalent to the action of the rotation-induced field $\mathbf{H}_B$ discussed above on the magnetic elements of the medium.

The first known attempt to verify this idea was made by John Perry, who mentioned that in his famous public lecture of 1890 [59]; it was not successful. Barnett began experimenting along this line in 1909; the first success was achieved in five years. These experiments and challenges encountered and overcome in their implementation are described in Barnett's review [55]; to feel the challenges, the rotation speed 100 *rps* (revolution per second) is equivalent of one fifteen-thousandth of the Earth magnetic field; the earth field was neutralized (as much as it was possible) and most experiments were performed in a wooden (nonmagnetic) building at night time to minimize disturbances of all kinds; the rotational speeds were in the range from 6 to 120 *rps*.

The measurements were performed with different ferromagnetic materials[15], such as iron, cobalt, nickel and many alloys. It was found that (a) the sign of γ is negative and (b) in all experiments the g factor is equal to 2, i.e., it is twice as big than that of the orbital motion. The error bar in first measurements was 12%, later it was reduced down to 2%. Therefore, Barnett came to the conclusion that magnetization of ferro- and paramagnets is due to electron spin[16] but not to its orbital motion. We will return to this really shocking fact in Sec. 2.3.3.

2.2.3 Rotation by magnetization (Einstein-de Haas effect)

The effect of rotation by magnetization discovered by Einstein and de Haas [54] is the inverse effect of magnetization by rotation.

A magnetic moment of a homogeneous ellipsoidal specimen suspended on a string in a uniform magnetic field $\mathbf{H}_0$ (e.g., a cylinder in a solenoid-like vertical coil) is

$$\mathbf{M} = \sum \mu = \sum \gamma \iota = \gamma \sum \iota = \gamma \mathbf{L}_m, \qquad (2.20)$$

where μ and ι are the magnetic moment and the angular momentum, respectively, of the magnetic elements composing the specimen, $\mathbf{L}_m$ is the total

---

[15] In the first successful experiment the specimen was an iron rod of nearly 1 m long and 7 cm in diameter; later much smaller specimens were used.

[16] Citing Barnett "Lorentz electron spinning on diameter".

angular momentum of these elements and the summation is taken over all of them.

If $\mathbf{M}$ is changed for $\Delta\mathbf{M}$, the angular momentum $\mathbf{L}_m$ changes for $\Delta\mathbf{L}_m = \Delta\mathbf{M}/\gamma$. Therefore, due to the momentum conservation, the specimen experiences recoil, i.e., its angular momentum $\mathbf{L}$ changes for $\Delta\mathbf{L} = -\Delta\mathbf{L}_m$. The torque $\boldsymbol{\tau} = d\mathbf{L}/dt$ acting on the specimen in this case is called *gyroscopic torque*, so that

$$\Delta\mathbf{L} = \int \boldsymbol{\tau}\,dt = -\frac{\Delta\mathbf{M}}{\gamma}. \tag{2.21}$$

If in the field H_0 the magnetization reaches its saturated value I_s, then at instantaneous reversal of $\mathbf{H}_0$ the specimen magnetic moment changes for $\Delta M = 2I_s V$. Accordingly, the resting specimen turns for an angle φ determined from the equation

$$\frac{D\varphi^2}{2} = \frac{Q\omega^2}{2}, \tag{2.22}$$

where D is the string torsional constant, Q is the specimen moment of inertia and $\omega = \Delta L/Q$ is the maximum angular velocity of the specimen.

Measuring this (ballistic) turning one can find γ:

$$\gamma = \frac{\Delta M}{(QD)^{1/2}}\frac{1}{\varphi} = \frac{2I_s V}{(QD)^{1/2}}\frac{1}{\varphi}. \tag{2.23}$$

The idea of an experiment of such kind goes back to Maxwell [1]. The first (unsuccessful) attempt to implement it was undertaken by Richardson[17] in 1908 [60]. Einstein and de Haas[18] mentioned but rejected this approach as insufficiently reliable[19]. Instead, Einstein and de Haas developed a resonance method, which allows one to magnify the effect.

If the current in the above coil changes sinusoidally with frequency ω, then the applied field and therefore the H field inside the cylindrical specimen is $H = H_0 \sin\omega t$, where H_0 now is the field amplitude. Hence, since the magnetization is proportional to H, the gyromagnetic torque τ is proportional to $\cos\omega t$. Then, after complete damping of natural oscillations, the specimen equation of motion (if τ is the only torque acting on the specimen) is

$$Q\ddot{\varphi} + P\dot{\varphi} + D\varphi = \tau_0 \cos\omega t, \tag{2.24}$$

where P is the damping coefficient and $\tau_0 = I_0 V\omega/\gamma = \chi V H_0\omega/\gamma$, where I_0 is the amplitude of oscillating magnetization.

Solution of this equation is well known (see, e.g., [44]): the specimen will oscillate as

$$\varphi = \varphi_0 \cos(\omega t + \delta), \tag{2.25}$$

---

[17]On that reason the effect of Einstein and de Haas sometime is referred to as the Richardson-Einstein-de Haas effect.

[18]They learned about Richardson's experiment after submitting their paper of 1915 [54]; nor did they know about Barnett's experiments.

[19]Significantly later the ballistic method was successfully used by number of authors; the measured g was close to 2 [55].

where δ is the phase difference between the φ (oscillations of the specimen) and τ (the driving torque); φ_0 and δ are determined by the system parameters Q, P and D.

At a definite frequency ω_r close to the system natural frequency $\omega_0 = \sqrt{D/Q}$ there will be a resonance at which the amplitude φ_0 takes on the maximum value φ_r. The resonance is the most intense, i.e., the amplitude φ_r reaches the maximum magnitude φ_∞[20], when $\delta = \pi/2$ or the oscillations of the specimen φ are behind the oscillations of the driving torque τ for a quarter of the period. In that case ω_r equals ω_0 and the amplitude is

$$\varphi_\infty = \frac{\tau_0}{P\omega} = \frac{I_0 V}{\gamma P}. \tag{2.26}$$

Hence, the measurement of the amplitude of oscillations at $\omega = \omega_0$ allows one to determine γ provided that $\delta = \pi/2$.

However, this scheme is inapplicable to the experiment of Einstein and de Haas. First of all, because in ferromagnetics (the materials with the greatest magnetization and therefore with the greatest gyroscopic torque) magnetization is not a linear function of H. The solution found by the authors is as follows. At sufficiently large amplitude of the current in the coil, the magnetization suddenly passes from one saturation value into the opposite one at a very short time interval when the current changes its direction. Respectively, the gyromagnetic torque, representing a delta-like function of different polarity, acts on the specimen at each half of the current period. Then this periodically changing torque with magnitude $2V I_s/\gamma$ is expanded into the Fourier series, from which the only first term needs to be considered at the resonance.

The authors did their best to minimize the disturbing torques due to residual terrestrial field (remained after its compensation), a possible misalignment of the cylindrical specimen and the applied field, the field inhomogeneity near the specimen ends, etc. They watched for the phase difference between φ and τ and found it to be close to the needed value $\pi/2$. The reported value of the g factor is unity with an estimated error about 10%, which is half of the value measured by Barnett.

In view of principally different physics behind $g = 2$ (Barnett) and $g = 1$ (Einstein and de Haas), obtained for the same material (in both cases the specimens were cylinders turned of soft iron) by different techniques, these works aroused great interest to the gyromagnetic effect. Many experiments with different ferromagnetic materials and advanced setups were performed in different laboratories. We already mentioned that the uncertainty of Barnett's measurements was refined from 12 to 2%. The ballistic and resonance techniques have also been greatly improved.

---

[20]The subscript ∞ means that this is the amplitude of the stable oscillations taking place when the natural oscillations of the system are completely damped out.

The most significant improvement of the resonance technique has been done by Coeterier and Scherrer [61] who used a flat-top current wave to power the magnet and forcefully set the phase difference δ between the gyromagnetic torque τ and the angular displacement φ equal to $\pi/2$. This was achieved using a photoelectric relay that reverses the polarity of the current. The relay was actuated by a narrow beam of light reflected from a mirror attached to a cylindrical specimen, which switched the relay when the specimen passed through its equilibrium position ($\varphi = 0$). This auto-phasing guarantees the sharpness of the resonance and automatically eliminates the effect of torques in quadrature with the gyromagnetic torque.

A large list of ferromagnetic materials has been investigated; some experiments were carried out with paramagnetics (see [55] for details and references). Except pyrrhotite in all studied materials the gyromagnetic ratio equals 2 within pretty small error bars.

2.2.4   The gyromagnetic effect in superconductors (Kikoin-Goobar effect)

The gyromagnetic effect in superconductors was for the first time measured by I. Kikoin and Goobar [45]. The principal motivations in setting up this experiment were to find out (1) if the gyromagnetic effect exists in diamagnetics and (2) whether it is possible in metals with infinite conductivity.

Indeed, the use of the gyroscopic model of Fig. 2.3 to diamagnetic atoms can be arguable, and therefore a priory it was not clear how and if the induced moment (always directed against the field) will react on the field change. Regular diamagnetics are unsuitable to verify this question due to too small magnetic susceptibility. In superconductors in the Meissner state χ is 5 orders of magnitude greater, which makes them appropriate for this goal. On the other hand, according to some theoretical interpretations (see, e.g., Sec. 3.2.2), the "superconducting electrons" are totally free, but, if so, then the specimen (i.e., the ionic lattice) may experience no recoil when the current due to these electrons changes.

The authors have chosen the resonance method of Einstein and de Haas in the version of Coeterier and Scherrer. The results obtained were communicated in [45]; a detailed report was later published in [62]. The principal challenges in measurement of the gyroscopic effect in superconductors are similar as those encountered with ferromagnetics. However in the given case these challenges are exacerbated by specifics of superconductivity (such as a possibility of the flux trap), comparatively small χ (50–100 times less than in ferromagnetics), and technical complexities caused by low temperatures.

An essential disturbing torque acting on the superconducting specimen in the chosen method is the torque arising due to residual terrestrial field and a minute misalignment between the field direction and the specimen axis in using cylindrical specimens. This torque was practically eliminated via (a) better

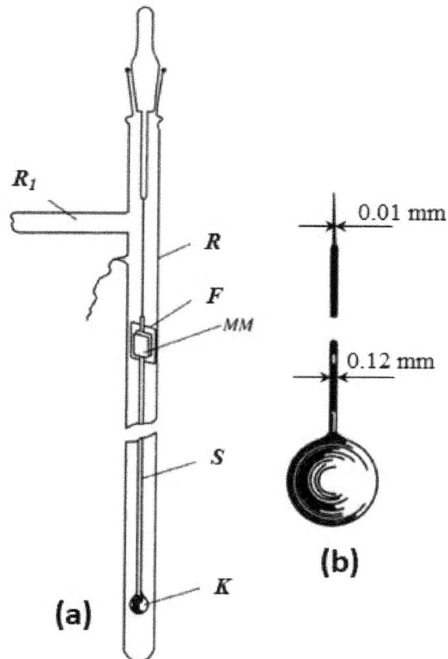

Figure 2.4 Schematics of the suspension system in the experiment of I. Kikoin and Goobar [45]. (a) Tube with the specimen. K is the spherical specimen; R is the glass tube silver plated on the inside; MM are two mirrors; S is a thin glass tube 0.12 mm in diameter and 45 cm in length glued to the specimen; F is a window; R_1 is a tube connected to a unit controlling helium pressure. (b) the specimen 3–4 mm in diameter; deviation of the thin tube axis from the specimen diameter did not exceed 10 μm. Adapted from [62].

than usual[21] compensation of the earth field (down to $5 \cdot 10^{-4}$ G) and (b) by using spherical specimens. The latter were lead balls 3–4 mm in diameter with a tolerance not exceeding 2–3 μm.

The suspension system with specimen is schematically shown in Fig. 2.4. The mirrors MM served for the optical control of the switching the polarity of the applied field H_0 and for reading the angular displacement φ via an image of a reflected light ray on the scale distanced from the system for $l = 170$ cm. The tube R_1 connects the system with a unit controlling helium pressure; changing the pressure changes the damping factor P. Initial helium pressure in R, equal to 10^{-2} Torr, provided thermal contact of the specimen with liquid helium in the cryostat.

---

[21]Typically, the residual field is on the order of $0.01 H_T$, where H_T is the terrestrial field.

Referring Eq. (2.26), the doubled amplitude of an oscillating light spot on the scale is

$$A_\infty = 2l\varphi_\infty = \frac{2l\tau_0}{P\omega} = \frac{2l\chi V}{\gamma P}\,H = \frac{l\chi VT}{\gamma Q\zeta}\cdot\frac{H_{01}}{1-\eta}, \qquad (2.27)$$

where $\chi = 1/4\pi$, T is the period of oscillations; $\zeta = PT/2Q$ is the logarithmic decrement; $H_{01} = 4H_0/\pi$ is the first harmonic of the oscillating applied field H_0 (the applied field has a shape of a squared wave with amplitude H_0); and η is the demagnetizing factor equal to $1/3$.

The measurements were conducted with the ZFC specimen at temperature 4.2 K. The critical field of lead at this temperature is about 470 Oe. In the experiments the field H_0 did not exceed 150 Oe.

Before each experiment the system was passed through multiple tests such as verification of the earth field compensation, verticality of the solenoid (the magnet), the absence of magnetic inclusions, etc. Preliminary rough measurements with an iron specimen yielded $g = 1.95$.

Nevertheless, the first experiments with lead were unsuccessful, which has been attributed to a trapped flux of the residual terrestrial field (10^{-4} G!). After replacement of the specimen for another one made of purer lead (provided by Kapitza), the results became stable. The stationary regime of resonant oscillations was achieved either at a gradual increase of the amplitude (swinging up starting with a small amplitude) or at its decrease (fading down starting with a deliberately large amplitude). Each experiment lasted about 10 hours. Results obtained in different modes and with different damping factors agreed with each other.

The reported value of the g factor is

$$g = 1 \pm 0.03$$

and the sign of γ is negative. Later the same result was obtained by Pry et al. with a tin sphere of an inch in diameter [63].

I. Kikoin concluded [62] that (1) the question about the gyromagnetic effect in superconductors has been answered positively, which apparently means that normally conducting diamagnetic bodies also possess gyromagnetic properties and, on the other hand, superconductors should exhibit the Barnett effect; (2) magnetization of superconductors is not due to electron spin, but closed electron currents. He noted that the g-factor of any orbital motion of spinless electrons is unity. He admits a possibility that closed electron currents could be circumferential surface currents (Foucault currents), "but then a difficulty arises regarding the mechanism of occurrence of these currents". Another possibility is that the diamagnetism of superconductors and the associated gyromagnetic effect are "caused by microscopic closed currents, but their origin is so far unknown". We remind the reader that this was written ten years prior to Cooper's paper on electron pairing in superconductors.

The reversed Kikoin-Goobar effect or magnetization of superconductors by rotation, often referred to as the London moment, was for the first time observed in 1964 by Hildebrandt [64]. We will discuss this effect in Sec. 5.2.10. Leaping ahead, the gyromagnetic effects in superconductors represent a direct proof of precession of Cooper pairs with mutually compensated electron spins, thereby confirming predictions of Edwin Hall [65] (see Sec. 3.1.1) and Isaak Kikoin [62] about microscopic origin of magnetization in these materials.

2.3   LANGEVIN'S THEORY OF MAGNETIZATION

In 1905 Langevin published his celebrated microscopic theory of magnetism in dia- and paramagnetics [66, 67, 68]. Coming from Lorentz' theory of electrons and ideas put forward by Ampère, Maxwell and J. J. Thomson, Langevin based his theory on the assumption that atoms are constituted by electrons of two kinds, negative and positive, moving in stable periodic orbits[22]; all atoms have spherical symmetry; in the field absence the magnetic moment of diamagnetic atoms is zero, while it is not zero and permanent in atoms of paramagnetic materials.

The theory turned out to be very successful. In particular, for diamagnetism it retains its leading position even today. In this section we discuss Langevin's theory with a primary attention to diamagnetism as it is relevant to superconductivity.

2.3.1   Diamagnetism

The Langevin formula for the magnetic susceptibility of diamagnetics follows from Eq. (2.19) for the induced magnetic moment caused by precession of the electron orbit. In an atom containing Z electrons the induced moment is

$$\boldsymbol{\mu}_a = -\frac{Ze^2\mathbf{H}}{4mc^2}r_i^2 \tag{2.28}$$

and therefore the susceptibility is

$$\chi = -\frac{Zne^2}{4mc^2}r_i^2, \tag{2.29}$$

where n is number of atoms per unit volume and r_i is rms radius of the induced circular currents in each atom, as discussed in Sec. 2.1.

Eq. (2.28) is the main formula of Langevin's theory, which he derived based on the Larmor theorem (traditional version). As can be seen from this formula, the theory is essentially reduced to the calculation of r_i. Langevin assumed

---

[22]Langevin gave no explanation of the stability of such an atom, in connection with which some authors point to the implicit quantum nature of Langevin's theory. A detailed discussion of the historical roots of Langevin's theory is available in [69].

that the polarized atom retains the spherical symmetry and the mean squared radius of all its electron orbits R_a^2 is the same as it was in zero field. Therefore,

$$\langle x^2 \rangle = \langle y^2 \rangle = \langle z^2 \rangle = \frac{1}{3} R_a^2. \tag{2.30}$$

Hence, if $\mathbf{H}$ is parallel to the z axis,

$$r_i^2 = \langle x^2 \rangle + \langle y^2 \rangle = \frac{2}{3} R_a^2 \tag{2.31}$$

and

$$\chi = -\frac{Zne^2}{6mc^2} R_a^2. \tag{2.32}$$

This is canonical Langevin's formula for the magnetic susceptibility of diamagnetics in the form given by Pauli [70]. Quantum mechanics yields an identical expression [49, 39, 23].

The radius R_a calculated from Eq. (2.31) using experimental data on χ is close in the order of magnitude to the atomic radius obtained from X-ray diffraction data [116]. Langevin's formula shows no temperature dependence of the diamagnetic susceptibility. This is consistent with experimental fact first revealed by Curie[23], which is valid for many but not all diamagnetics.

2.3.2   Refinements and additions to the theory of diamagnetism

Perhaps the reader feels some discomfort from Langevin's assumption that the polarized diamagnetic atom retains the spherical symmetry, while the induced current has 2D symmetry (see Fig. 2.2b). Let us consider this issue and also obtain some relationships which we will need later.

1. *Radius of the induced current* r_i. It has long been noticed (see, e.g., [71]) that the law of temperature independence of the magnetic susceptibility is violated in some diamagnetic materials. Citing Van Vleck [23], "the worst offender is bismuth", whose susceptibility is constant above melting temperature, but rises sharply by a factor of nearly 30 at solidification; on top of that χ in crystalline bismuth is strongly anisotropic. Ehrenfest hypothesized that it can be due to the orbits spanning multiple atomic units [72]. Raman [73] supported Ehrenfest's argumentation and extended it to graphite [perpendicular to the hexagonal axis the magnetic susceptibility of graphite is nearly the same as in diamond (about $-5 \cdot 10^{-7}$), but along this axis $\chi \approx -2 \cdot 10^{-5}$ [71] and close to $-3 \cdot 10^{-5}$ in pyrolytic graphite]. On the other hand, if Ehrenfest and Raman are right, then, in the framework of the spherical symmetry of the polarized atom, the anisotropy of χ means that the atomic radius R_a depends on the direction of the magnetic field.

---

[23]Pierre Curie was the professor of Langevin.

These facts cast doubt on the validity of the spherical approximation (Eq. (2.30)), but do not raise questions in the interpretation based on Lorentz's force. Indeed, apart from pure intuitive, which can follow from, e.g., Fig. 2.1b, there is no any justification for Eq. (2.30) and, hence, to Eq. (2.31) or to the original Langevin's formula with $r_i^2 = R_a^2/3$ [67]. On the other hand, as already noted, the inconsistency of the spherical approximation follows from the fact that the induced current lies strictly in the transverse to $\mathbf{H}$ plane and therefore a polarized atom cannot maintain the spherical symmetry it had in the field absence.

Below we will see that in superconductors r_i^2 is reversely proportional to the number density of Cooper pairs. This suggests that r_i is not a property of a single microscopic unit, but rather a collective property of the medium. If so, one can expect that in normal diamagnetics r_i depends on geometry of the atomic structure in the crystalline plane transverse to $\mathbf{H}$. Recognition of the fact that r_i and R_a are different, although undoubtedly related quantities, can explain the observed susceptibility anomalies in graphite, bismuth and other anisotropic diamagnetics [71].

2. *Entropy.* The absence of residual magnetization and the temperature independence of the magnetic susceptibility in isotropic diamagnetics indicate that the magnetic field does not change the magnetic order; i.e., a magnetic part of entropy S_m in diamagnetic specimens is zero (see problem 1.2). The latter in its turn means that all induced circular currents are in phase and therefore the induced velocities $\mathbf{v}_i$ (Eq. (2.14)) are the same (i.e., both in magnitude and direction) in all atoms (or other microscopic constituents) of the given specimen[24]. Herewith, the absolute value of the phase is undefined and can be different in otherwise identical specimens.

3. *Magnetic energy of a cylindrical specimen.* Imagine a long diamagnetic cylindrical specimen in a solenoid-like magnet wound immediately on the specimen; the demagnetizing factor of such a specimen η is zero. Now turn on the magnet power supply and set the field equal to $\mathbf{H}_0$. In the chosen geometry the field intensity in the specimen $\mathbf{H} = \mathbf{H}_0$ and therefore the magnetic moment acquired by the specimen $\mathbf{M} = \chi V \mathbf{H} = \chi V \mathbf{H}_0$. Herewith, since magnetization does not change temperature of diamagnetics, the change of the specimen internal energy equals its magnetic energy E_m, defined as

$$E_m = -\int_0^{H_0} \mathbf{M} d\mathbf{H}_0. \tag{2.33}$$

The energy E_m represents the work required to transport the specimen from zero field to the field $\mathbf{H}_0$ or this is the work done by the magnet power

---

[24]In terms of quantum mechanics, the phase (angle) of the electron circular motion corresponds to the phase of its wavefunction. The different phases of the wavefunctions of individual electrons means the presence of a phase gradient in the wavefunction of electron ensemble. If this were so, it would lead to the appearance of currents other than orbital ones, which is impossible in thermodynamically equilibrium systems.

supply to magnetize the specimen minus the work to establish $\mathbf{H}_0$ without the specimen [16].

Next, since the magnetized specimen of the cylindrical geometry does not create the field outside itself, by virtue of the energy conservation E_m equals the change of kinetic energy of electrons ΔT induced by the magnetic field. Let us look what it is equal to.

The change of kinetic energy of the orbiting electrons in an atom ΔT_a is

$$\Delta T_a = \frac{m}{2} \sum_a \left[(\mathbf{v}_0 + \mathbf{v}_i)^2 - \mathbf{v}_0^2 \right] = m \sum_a \mathbf{v}_0 \cdot \mathbf{v}_i + \sum_a \frac{m\mathbf{v}_i^2}{2}, \qquad (2.34)$$

where the sum is taken over all electrons in the atom.

The first term in the right hand side vanishes due to the condition of Eq. (2.10). Thus, the change of kinetic energy of electrons in the polarized atom equals the sum of kinetic energies of their motion induced by magnetic field; i.e., the change of kinetic energy per atom is

$$\Delta T_a = \sum_a \frac{m\mathbf{v}_i^2}{2} = Z \frac{mv_a^2}{2}, \qquad (2.35)$$

where v_a^2 is the mean square of induced linear velocities in the atom, which is equal to $(or_i)^2$ because o and $\mathbf{r}_i$ are perpendicular to each other.

Next, using Eqs. (2.1), (2.29) and the fact that $\mathbf{H} = \mathbf{H}_0$ in our specimen, for the change of kinetic energy of electrons per unit volume ΔT_v we write

$$\Delta T_v = \frac{Znm}{2} \frac{e^2}{4m^2c^2} H^2 r_i^2 = -\left(-\frac{Zne^2}{4mc^2} r_i^2 \right) \frac{H^2}{2} =$$

$$-\frac{1}{2}\chi H_0^2 = -\frac{1}{V} \int_0^{H_0} \mathbf{M} d\mathbf{H}_0 = \frac{E_m}{V}. \qquad (2.36)$$

Thus, we have shown that E_m of a diamagnetic specimen of the cylindrical geometry equals kinetic energy of the field induced motion of the bound electrons. This statement reflects the law of energy conservation and therefore its fulfillment is mandatory for any diamagnetic specimen with $\eta = 0$.

2.3.3 Paramagnetism

The basic assumptions of the Langevin theory are: each atom in a paramagnetic specimen possesses a fixed (independent of the field and temperature) magnetic moment, which do not interact with the moments of other atoms. The medium is assumed isotropic and atoms are subject to thermal agitation.

The Langevin formula (more correctly, similar to it) for the magnetization of paramagnetics can be derived as follows.

The magnetic moment of each microscopic element consists of the momenta due to orbital (orbital moment $\boldsymbol{\mu}_{or}$) and spin (spin moment $\boldsymbol{\mu}_s$) angular momentums of electrons. The magnitude of $\boldsymbol{\mu}_s$ equals $\gamma\hbar/2 = \hbar ge/4mc = \hbar e/2mc = \mu_B$, where μ_B is the Bohr magneton. For simplicity, let us consider a non-conducting specimen consisting of mono-atomic molecules whose magnetic moment is due to uncompensated μ_{or} and μ_s of one of its electrons.

The specimen magnetic moment is

$$\mathbf{M} = \sum \mu_{or} + \sum \mu_s, \qquad (2.37)$$

where summation is taken over all atoms.

In the field absence both terms are zero due to 3D symmetry of the medium. According to the results of measurements of the gyromagnetic effect (Sec. 2.2), the dominant contribution to the magnetization of paramagnetics is made by the spin moments. Therefore, the orbital term in Eq. (2.36) can be neglected[25]. As we know from the Larmor theorem, in the field $\mathbf{H}$ each atom rotates with the Larmor frequency $\mathbf{o}$ about direction of $\mathbf{H}$. Hereby, in the rotating coordinate system S' the field acting on electrons is zeroed by the Barnett field (see Sec. 2.1). This means that spins of all orbiting electrons, including the uncompensated one, are totally free; i.e., they *do not precess*.

So, we have a situation similar to that depicted in Fig. 2.3, where now AA is the spin axis (the moment μ_s) but without the springs. Therefore, depending on the spinning direction, the moments μ_s turn orienting either up (parallel) or down (antiparallel) to $\mathbf{H}$ (the angle θ becomes either 0 or π). Correspondingly, those spins which were compensated remain compensated, whereas the uncompensated spins are oriented along (up or down) the field. At the same time, nothing is changed in the mutual orientation of the orbital moments because all of them precess synchronously. Hence, in the field an average atom acquires a magnetic moment equal to μ_s orientated along (up or down) the field.

Now, recalling that the field $\mathbf{H}$ can be treated as a potential field (see Sec. 1.2.6), we can assign to each atom a potential energy[26] $U = -\mu_s \mathbf{H}$. The latter means that atoms whose magnetic moment $\mu(= \mu_s)$ is parallel to $\mathbf{H}$ have lower energy than those whose moment is directed against the field. On the other hand, since our specimen is in thermodynamic equilibrium, we can apply the Boltzmann theorem.

We note in passing, that the energy needed to flip μ_s equals $2U = 2\mu_s H = (\hbar e/mc)H$ or the linear frequency of the quantum of electromagnetic radiation of this energy is $\nu = (e/2\pi mc)H = 2.8 \cdot 10^6 H$. This is the condition of the electron paramagnetic resonance formulated in a celebrated paper of Yevgeny Zavoisky in which the discovery of the EPR (also called ESR, the electron spin resonance) was reported [74].

The Boltzmann theorem states that distribution of particles (atoms, ions, electric dipoles, etc) constituting a thermodynamically equilibrium statistical system in the presence of a conservative field differs from their distribution in

---

[25] This is the so-called frozen or quenched orbital momentums approximation.

[26] Of course, no potential energy exists in a magnetic field due to the solenoidal (non-conservative) nature of the latter. One can show that $-\mathbf{MH} = \Delta T$, where ΔT is the change of kinetic energy of the current carriers creating the magnetic moment $\mathbf{M}$ [18].

the field absence by the factor $exp(-U/k_BT)$, where U is the potential energy of a particle in this field, k_B is the Boltzmann constant and T is temperature.

So, denoting number of atoms of our specimen whose moment is parallel and antiparallel to the field as N^+ and N^-, respectively, and the total number of atoms as $N = N^+ + N^-$, we write

$$\frac{N^+}{N} = \frac{e^{\mu_s H/k_B T}}{e^{\mu_s H/k_B T} + e^{-\mu_s H/k_B T}} \tag{2.38}$$

and

$$\frac{N^-}{N} = \frac{e^{-\mu_s H/k_B T}}{e^{\mu_s H/k_B T} + e^{-\mu_s H/k_B T}}. \tag{2.39}$$

Accordingly, the expression for magnetization has the form

$$I = \frac{\mu_s(N^+ - N^-)}{V} = n\mu_s \frac{e^x - e^{-x}}{e^x + e^{-x}} = n\mu_s \tanh x, \tag{2.40}$$

where $n = N/V$ is number density of atoms and $x = \mu_s H/k_B T$.

This is a modified Langevin's formula for magnetization of paramagnetics, which he derived considering temperature agitation of precessing orbital magnetic momenta. The original Langevin's formula is

$$I = n\mu_{or}\left(\coth x - \frac{1}{x}\right). \tag{2.41}$$

Comparison of magnetizations following from Eqs. (2.40) and (2.41) normalized with respect to the saturated value I_s (as it was presented in [67]) is shown in Fig. 2.5. The formula Eq. (2.40) is identical to a quantum-mechanical

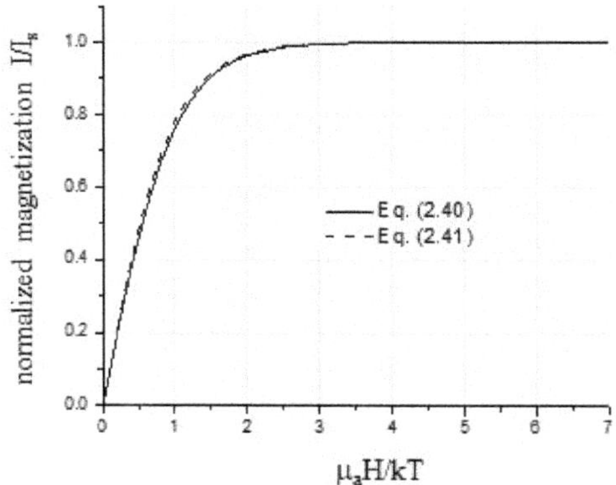

Figure 2.5 Normalized magnetization calculated from Eq. (2.40) and (2.41). I_s is the saturated magnetization. μ_a is the atomic magnetic moment, which is equal to μ_s in Eq. (2.40) and μ_{or} in Eq. (2.41).

formula for a two-level spin system [40, 49]. Experimental data (see, e.g., [75]) unambiguously support the quantum-mechanical formula with the frozen orbital momenta. A quantum-mechanical interpretation of the paramagnetism is given in the monographs of Van Vleck [23] and Bloch [39], as well as in textbooks, e.g., [40, 41, 47].

As seen from Fig. 2.5., at low fields ($x \ll 1$) both formulae yield linear dependence of I vs. H. Correspondingly, the susceptibility in both cases is

$$\chi = \frac{C}{T},\tag{2.42}$$

where C is a constant equal to $n\mu_s^2/k_B$ according to Eq. (2.40) and $n\mu_{or}^2/3k_B$ according to Eq. (2.41).

Eq. (2.42) is a law discovered by Curie ten years prior the development of the theory. For a strong field ($x \gg 1$), the theory predicts saturation, not yet known in 1905, but brilliantly confirmed later.

It remains for us to explain the absence of contribution of the orbital momenta to the magnetization. This can be done as follows.

As we have seen, in the field averaged atoms have a magnetic moment equal to $\boldsymbol{\mu}_s$, which are free and therefore react on the thermal agitation in accord to the Boltzmann theorem. On the other hand, each (now really each, not averaged) atom also possesses a magnetic moment due to the uncompensated orbital magnetic moment of one of its electrons. Then why the orbital momenta do not contribute to magnetization, as follows from a huge amount of experimental data on the gyromagnetic effect, magnetization and EPR? In other words, why the orbital momenta are "frozen"? The answer is: because (1) the orbital momenta cannot change their orientation continuously due to the spacial quantization and (2) they are not free [46]. Indeed, the orbital momenta of individual electrons within the atom are rigidly bound to each other and with the momenta of other atoms via interatomic bounding (recall that all magnetic momenta in the specimen precess synchronously). These facts make it much harder to react on the thermal agitation for the orbital momenta than that for the spins. In 1905, nothing foreshadowed the existence of either spatial quantization or electron spin. Therefore, Langevin constructed his theory assuming a continuous change in the orientation of $\boldsymbol{\mu}_{or}$. The erroneousness of such an assumption is the reason of the quantitative discrepancy of the Langevin theory with experiment [75] and a conflict with the Third law (Nernst theorem) [39], which was yet to be formulated in 1905.

CHAPTER 3

SUPERCONDUCTIVITY. HISTORY OF DEVELOPMENT

3.1 DISCOVERY AND PRINCIPAL PROPERTIES

3.1.1 Zero resistance

On 8th of April 1911 Kamerlingh Onnes made an entry in his notebook, which in English translation reads "Mercury practically zero". It was the first observation of superconductivity, a phenomenon that has topped the list of researches in condensed matter physics for more than a century. In the fall of same year, at the First Solvay Conference, Onnes reported on the results of this and two more experiments [4], where, in particular, he presented which has become historical graph shown in Fig. 3.1[1].

Events in Leiden were developing at a tremendous pace: nearly each experiment with liquid helium[2] brought a new discovery [76, 77]. Already in the subsequent year it was found that lead and tin also superconduct below about 6 and 3.8 K, respectively. 1914 was marked by the discovery of the critical magnetic field, critical current and the first demonstrations of a so-called persistent current set up by induction in a short-circuited lead coil and

---

[1]In the entrance hall to the Kamerlingh Onnes Building at Leiden University there is a plaque: "On 8 April 1911, in this building, Professor Heike Kamerlingh Onnes and his collaborators, Cornelis Dorsman, Gerrit Jan Flim and Gilles Holst, discovered superconductivity. They observed that the resistance of mercury approached "practically zero" as its temperature was lowered to 3 kelvins. Today, superconductivity makes many electrical technologies possible, including Magnetic Resonance Imaging (MRI) and high-energy particle accelerators".

[2]Helium was first liquefied in the laboratory of Kamerlingh Onnes in 1908. Until 1922, this was the only laboratory possessing liquid helium.

DOI: 10.1201/9781003355786-3

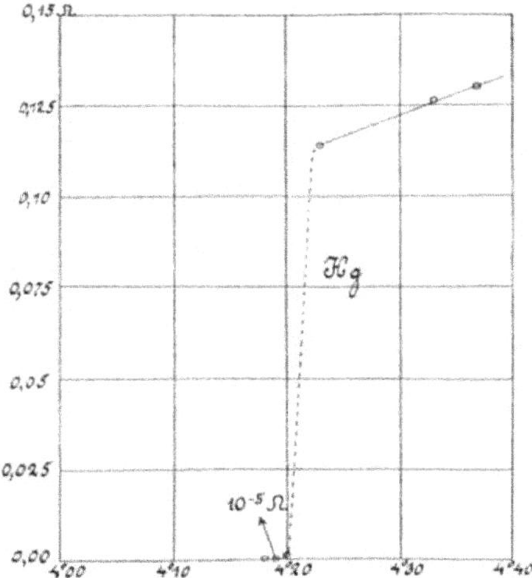

Figure 3.1 Resistance of mercury wire vs temperature (K) measured in the Lab of Kamerlingh Onnes in October 1911 [4]. The resistance 0.125 Ω constitutes 1/500 of its value at room temperature.

ring. Soon the list of superconductors was replenished with thallium and indium. Ever since the amount of superconductors is constantly growing. In the 1960s, the number of known superconductors was close to 1000 [78]. Nowadays, it is hardly possible to enumerate them all, which clearly indicates that superconductivity is a widespread phenomenon.

Returning to Leiden, here Onnes focused his research on superconductivity at a "residual micro-resistivity" of the S state. In his last full-scale project, reported in 1924 at the 4th Solvay Conference[3], the upper limit of the resistivity of superconducting lead was estimated as 10^{-12} relative to its value at room temperature [79]. At the same time, this project revealed a new mysterious property of superconductors.

In brief, the conducted experiments were as follows. A lead ring was suspended on a torsion spring within a slightly larger fixed lead ring and the assembly was immersed in liquid helium. Initially, when the rings were in the same plane, currents in both of them were induced by an external magnetic field transverse to the rings. Then, by twisting the torsion spring, the suspended ring was turned in the horizontal plane by $30°$ and a change of its angular position (proportional to the change in current) was monitored on a scale with an image of a narrow light beam reflected from a mirror attached to the spring. After about 20 minutes of relaxation, the ring position got stabilized and remained still over all the rest time of observation (about 6 hours). At the next stage, the suspended ring was replaced by a lead shell deposited

---

[3]The report was presented by Keesom in connection with the illness of Onnes.

on a spherical glass bulb; result was the same. After that Onnes and assisted him Tuyn used a ring made of soldered together 24 alternative sectors of tin and lead[4]; each sector represented a band of tin or lead wrapped around a ring of ivory. It was expected that the induced current would decay, however, the result was the same as with the continuous ring and the shell: after the relaxation, the light image on the scale stood still[5]. Finally, Onnes and Tuyn just cut the ring: no change in the current was registered[6]!

Apart from the estimate of micro-resistivity, from these experiments Onnes concluded that the Hall effect in superconductors is absent, which was consistent with a previously observed disappearance (a sharp drop down to undetectable level) of the Hall voltage in lead and tin samples at the transition to the S state [82]. This conclusion was based on an assumption that the persistent currents in the rings and shell run over circumferential surface paths which, being formed at the N/S transition, stay unchanged. Lorentz supported this assumption [83]; Einstein also endorsed it, but in a postscript of a paper on this matter he admits that this "speculation (which by the way is not new)" contradicts Onnes' results with the sectorial ring [84].

Later Hall, who listened Onnes' report in Brussels, commented on it before the US Academy of Science [65]. There he raised, as he said, a "heretical question". Quote: "Is there conclusive evidence that the persistent currents which Onnes and others have observed are anything more than the aggregate of microscopic electric whirls within the metal? Is there conclusive evidence that the persistent current which is ordinarily assumed to be circumferential within a supraconductive ring or shell is really circumferential?" Note that Hall's hypothesis of the current whirls can explain all observations of Onnes and Tuyn, provided the whirls can move.

Summarizing, the first about 15 years the research on superconductivity was mostly concentrated on studies of the total (macroscopic) superconducting currents. These currents, astonishingly resembling the currents in an ideal or perfect conductor discussed by Maxwell [1], were (1) transport currents maintained by an applied e.m.f. in a circuit with a superconducting section; and (2) the currents induced in multiply connected superconducting samples (short-circuited coils and rings).

The resistivity of real superconductors with the total current is vanishingly small, but not zero, as was expected by Onnes. The "micro-resistivity" is associated with magnetic properties of superconductors, specifically with different kind of the flux flows, or motion of normal domains with multiple and single flux-quanta [85]. The motion of normal domains in a type I superconductor is demonstrated in Video 1 in the Support Materials.

---

[4]The idea of this experiment was suggested by Einstein [80].

[5]Herewith, the induced magnetic moment was lesser than that in continuous ring [81].

[6]Quoting Onnes [79]: "Nous avions pensé que nous trouverions un courant qui prendrait un certain temps à s'eteindre. Mais l'experience a montré que des courants continuaient a circuler dans l'anneau et, lorsque l'experience fut repetee avec l'anneau coupe, celuici montra le méme moment magnétique".

Introducing impurities (so-called pinning centers or merely pins), this motion can be significantly suppressed but cannot be completely eliminated. A historical overview of this subject with a representative list of references is available in [86]. The decay time of the total current in a closed superconducting circuit is determined by the field, specifics of material and design of the circuit. For instance, as was measured using NMR by File and Mills [87], the current in a superconducting solenoid made of Nb-25%Zn with an initial field 2.1 kG decays with the rate of the order of 10^{-6} G/h, which corresponds to the total decay time about 100 thousand years[7] and an equivalent resistivity of the used wire on the order of 10^{-22} ohm·cm.

3.1.2 Zero entropy

Already at the first Onnes' presentation at the Solvay conference in 1911 Langevin suggested that superconductivity could be an unknown before thermodynamic phase of matter associated with an alteration of properties of conduction electrons. A similar proposal was put forward again by Langevin and Bridgman at the fourth Solvay Conference in 1924 when discussing the Onnes report presented by Keesom [89]. However, Lorentz and Keesom objected to the applicability of thermodynamics on the ground that, as was shown by Maxwell [1], magnetic properties of a resistanceless metal are irreversible, in full accordance with experiments on persistent current[8].

For quite a while, despite the search for an effect of superconductivity in other properties, it looked so that zero resistivity is the only property distinguishing metals in the S and N states[9]. The first breakthrough occurred in 1927, when Walther Meissner discovered a disappearance of a thermo-e.m.f. $\mathcal{E}$ (Seebeck voltage) in the S state [5]. This result was confirmed in studies of other researchers, and it was found that all equilibrium thermoelectric effects (Seebeck, Peltier and Thomson effects) vanish in the S state [15].

The absence of the thermoelectric effect made possible to establish an absolute scale for thermoelectric power or Seebeck coefficient ($\mathcal{E}$ per 1 K) of a metal via measuring $\mathcal{E}$ with respect to a superconductor, as was demonstrated by Borelius et al. [90]. This work marked the first practical application of superconductivity. Subsequently, using data on the Thomson effect, the absolute thermoelectric power was tabulated for many metals up to 2000 K [91]. Fig. 3.2 shows the Seebeck coefficient of lead measured with respect to the superconducting Nb_3Sn alloy [92]. Interesting to note the rapid change in the Seebeck

---

[7] In a more detailed report, the measured decay time varied in different runs from 4 to 29 thousand years [88].

[8] Interesting, that during this discussion, Keesom introduced a thermodynamic formula for the latent heat of the S/N transition.

[9] The drop in the Hall voltage [82] could also be an effect caused by the puzzling "superconducting electrons". Since these electrons, as seemed, are insensible to the presence of neighbors, then they can be also insensible to the transverse forces exerted by the field [83].

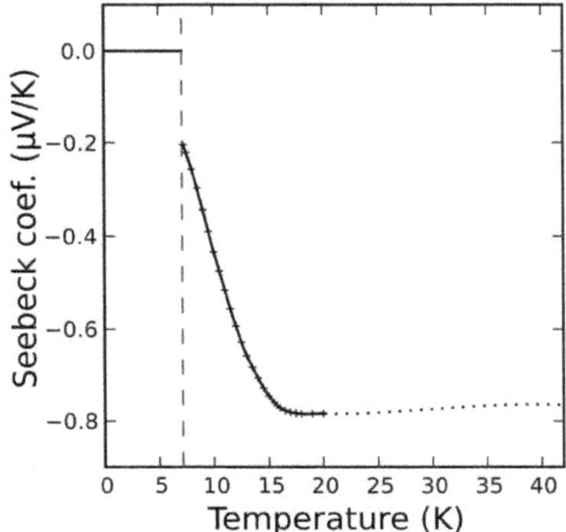

Figure 3.2 Absolute Seebeck coefficient of lead. The dashed line marks the critical temperature T_c. The dotted line is calculation using Thomson's relation. After Christian et al. [92]; reprinted with permission from the Royal Society of London.

coefficient above the critical temperature, which precedes a discontinuous drop to zero at T_c; such a dependence was anticipated by Shoenberg [15].

On the first glance, the significant attention to the thermoelectric effects (Shoenberg reviewed 10 experimental works performed before 1952) may look strange because the existence of this effect seems inconsistent with the resistanceless state of the metal: the non-zero termo-e.m.f. would lead to an infinite current and therefore to restoration of the N state. However, these studies were profoundly meaningful on the following reasons.

(1) By definition, the Seebeck voltage is the electromotive force developing in metals due to the transport of entropy without involvement of the electrical transport [93, 41]. Accordingly, the infinite conductivity a priory does not exclude the Seebeck effect[10]. (2) Zero thermo-e.m.f. means absence of the entropy transport and hence no temperature dependence of entropy in the superconducting fraction of the conduction electrons. (3) Taking into account the Third law (Nernst theorem), the latter condition means that entropy of the superconducting electrons is zero over entire temperature and field range of the S state.

Zero entropy of the superconducting electrons is the base postulate of the two-fluid model of Gorter and Casimir [6] discussed in Sec. 3.2.1. If this postulate is correct, then superconductivity can represent an unknown before thermodynamic state of matter characterized by complete ordering of the

---

[10]Ginzburg in 1944 argued that the Seebeck effect exists in superconductors [94]. Later he repeatedly insisted on this statement. Ginzburg's argumentation is based on the London theory discussed below.

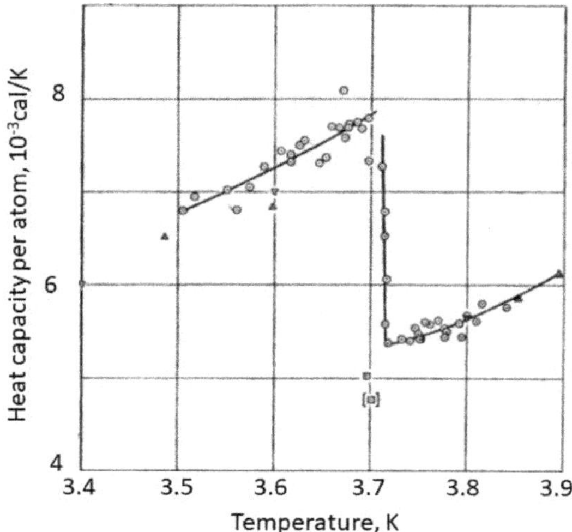

Figure 3.3 Atomic heat capacity of tin near the critical temperature $T_c = 3.72$ reported by Keesom and Kok [97].

superconducting electrons. Naturally, it was hard to accept this fact in view of the already mentioned irreversibility of the magnetic properties of perfect conductors.

3.1.3   Zero induction (Meissner effect)

By the late 20s and early 30s, evidences began to emerge indicating that superconductors differ from the hypothetical perfect conductors. For example, de Haas and Bremmer revealed that thermal conductivity of Sn and Pb in the S state is poorer than that in the N state [95], and therefore the Wiedemann–Franz law breakes down. Note, that this fact is consistent with zero entropy of the superconducting electrons.

In 1932 Keesom with co-workers performed a series of extremely challenging measurements of heat capacity in superconducting (Sn and Tl) and non-superconducting (Ag and Zn) metals[11] at zero field. They found that heat capacity of Sn and Tl experiences a sharp jump at the critical temperature [96, 97, 98]. The data reported in [97] are shown in Fig. 3.3.

This was a revolutionary discovery indicating the presence of a phase transition. Hence, superconductivity can be indeed an unknown before thermodynamic phase of matter. Results of Keesom's group served as a powerful call to study magnetic properties, and corresponding experiments were set up in

---

[11]Zn superconducts at temperature below 0.855 K, which was not achievable at the time.

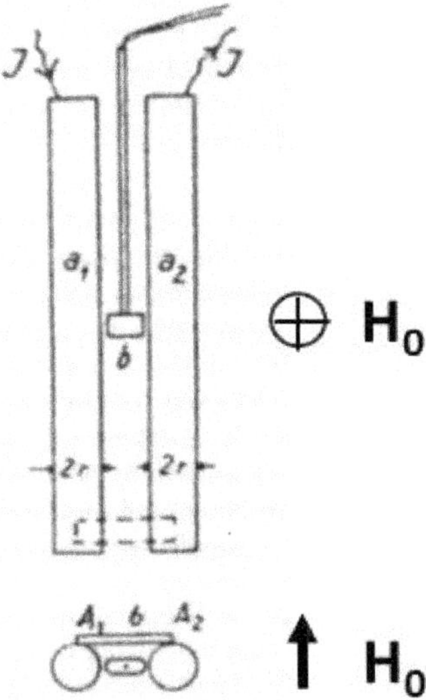

Figure 3.4   The experimental setup of Meissner and Ochsenfeld with two cylindrical specimens (side and top views) used in the second and fourth arrangements. a_1 and a_2 are the specimens of radius r, b is the search coil. J and A_1-A_2 are the current and the connecting bridge, respectively. H_0 is the applied field directed normal to the specimens (adapted from [100]).

four out of five laboratories possessing liquid helium at that time. The first convincing results were obtained by Meissner and Ochsenfeld in Berlin [7] and Ryabinin and Shubnikov in Kharkiv [8]. These works mark a turning point in studies of superconductivity: the discovery of an effect, which has become known as the Meissner effect. The historical experiments of Meissner and Ochsenfeld, and Ryabinin and Shubnikov are briefly reviewed below.

Meissner and Ochsenfeld [7] reported on measurements of the magnetic field in the vicinity and inside of superconductors in four arrangements[12]. The specimens were single-crystalline tin and poly-crystalline lead cylinders (130–140 mm in length and close to 10 mm in diameter [81]) placed vertically in the horizontally applied uniform magnetic field $\mathbf{H_0}$ as shown in Fig. 3.4. The measurements were performed using a small search coil connected to a ballistic galvanometer. The coil could be moved round the specimen and rotated in the horizontal plane without opening the cryostat. The current induced at turning the coil for $180°$ is proportional to the coil cross-sectional area and the induction B in the coil location, thus allowing to find an averaged field over the coil volume.

---

[12]An active promoter of these experiments was Max von Laue [99]; von Laue and Moglich performed calculations used by Meissner and Ochsenfeld.

In the first arrangement the field distribution near one specimen was measured after it was cooled below T_c in $H_0 \approx 5$ Oe[13]. According to the Faraday law, the field should stay undisturbed since no e.m.f. was induced. However, it turned out that below T_c the field near the cylinder changed almost to that which would be expected if μ_m in the S state is zero or $\chi = -1/4\pi$, the minimum permissible value.

In the second arrangement two parallel tin or lead cylinders were cooled in the same transversely applied field. It was found that below T_c the field between the tin specimens increased for a factor 1.70; for the lead cylinders this factor was 1.77. The increase factor for the field in the location of the search coil calculated coming from zero permeability of the S state was 1.77. These data support the statement above about zero permeability of the sample material in the S state. Note that no eddy currents can be induced under conditions of these experiments unless the Faraday law is broken.

In the third arrangement the sample was a hollow lead cylinder (a tube with the wall 2 mm thick). It was again cooled in the same field as before and B was measured inside the tube and adjacent to it outside[14]. It was found that the outer field changed in about the same way as it was for the solid cylinder. The inner field changed also: it increased for about 5%. The authors were not able to establish if the field inside remained uniform. On switching off the applied field keeping the sample superconducting, the field inside remained unchanged; at the same time the field outside decreased but it did not become zero.

In regard of this arrangement, the authors noted that their observations may look inconsistent with the statement about zero permeability. They suggested that these results can be explained in terms of microscopic or macroscopic currents in the superconductor assuming that $\mu_m = 1$ for the current-free (i.e., "normal") regions. Now we understand that due to a non-ellipsoidal shape, the tube sample was in neither one of the equilibrium superconducting states, the field passes through it via irregular N domains, and the flux is trapped when H_0 is switched off [16]. This means that there is no contradiction between results of the first two arrangements with those of the third one, and that Meissner and Ochsenfeld were exactly right in their interpretation.

After all, in the fourth arrangement two tin cylinders used in the second arrangement were connected end-to-end in series and a dc current of about 5 A was introduced through the other ends (see Fig. 3.4). It was found that below T_c the field between the specimens increased, although the current was kept unchanged. The measured field was about the same regardless whether the current was introduced before or after the specimen temperature passed through T_c, and in both cases the field readings were larger than that calculated assuming the surface superconducting current [81].

---

[13]Detailed results obtained with one specimen were reported in [101].
[14]Details were reported in [102].

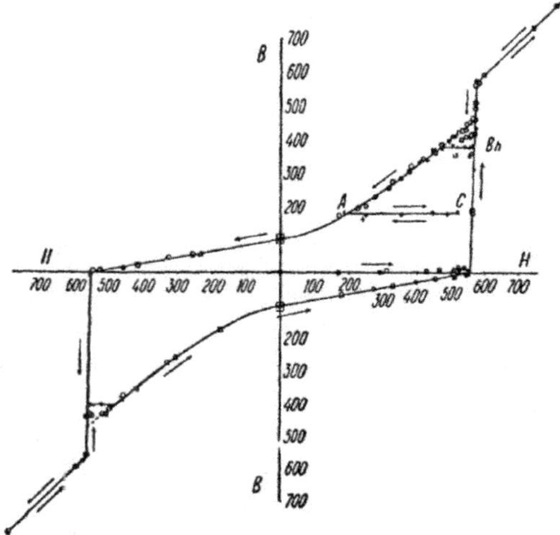

Figure 3.5 Induction B as a function of the field strength H in a cylindrical specimen measured by Ryabinin and Shubnikov [8]. The arrows indicate direction of variation of the applied field H_0. In the given geometry $H = H_0$. Reprinted from [104] with permission the American Physical Society.

The main conclusion of Meissner and Ochsenfeld was that B in their specimens below T_c is always zero. However not all researchers agreed with that (see, e.g., [103]). The experiment of Rjabinin and Shubnikov removed all doubts (although not at once).

Rjabinin and Shubnikov [8] attacked the same problem via measuring the magnetic moment $\mathbf{M}$ of a superconducting lead rod (5 mm in diameter and 50 mm long) at constant temperature 4.2 K vs $\mathbf{H}_0$ applied parallel to the rod axis. Two methods were used, which are similar to those employed in contemporary ac and dc magnetometry. In the first method the change $\Delta\mathbf{M}$ was determined by measuring the current induced in a pickup coil tightly wound around the middle of motionless sample at a sudden change of the applied field in small steps $\Delta\mathbf{H}_0$. In the second method the current in the pickup coil was induced by quickly removing the specimen away from the coil at fixed $\mathbf{H}_0$. Results obtained by both methods were consistent with each other but the second method appeared to be more reliable. So, the discussion was mainly based on the results obtained via dc measurements, which are shown in Fig. 3.5.

It was found that (a) when the sample was first magnetized after cooling in zero field, B and μ_m were zero at $H \leq H_c$; in a narrow field interval near H_c the induction B rapidly changed to a magnitude equal to that in the normal metal; at $H > H_c$ $\mu_m = 1$. (b) At decreasing H, $B = H$ until H reached its critical value; at H close to H_c, the induction experienced a sudden jump down, but it did not become zero; with a further decrease of the field B was also decreasing; at $H = 0$ there remained a residual induction close to 18% of the maximum B at $H = H_c$. The observed magnetization *loop* was reproducible.

The authors wrote "the actual fact that a jump takes place in the induction in falling field strengths we are incline to ascribe to the formation of a new phase with $B = 0$". The incomplete reversibility of the data obtained was attributed to imperfections of the sample material. As now well known, the interpretation of Rjabinin and Shubnikov was correct.

The experimental results of Meissner and Ochsenfeld plus Rjabinin and Shubnikov, confirmed in experiments of Tarr and Wilhelm [105], and Mendelssohn and Babbitt [106] once and for all changed the landscape of superconductivity. Specifically, it was established that at definite conditions a superconductor can be found in a reversible state, referred to as the Meissner state (MS), which is characterized by zero induction simultaneously with zero resistivity and zero entropy. The fact of reversibility means that the MS is an equilibrium thermodynamic state and therefore it obeys the Law of symmetry of the time reversal and, accordingly, requirements of the impossibility of the total current(s) and mutual compensation of all currents in the specimen bulk (see Sec. 1.2.1).

Important to stress, that the Meissner effect and, accordingly, the MS occur only in sufficiently pure and massive singly connected ellipsoidal specimens when the intensity H inside them lesser H_{c1} or the applied field H_0 lesser $H_{c1}(1 - \eta)$. Here H_{c1} is the lower critical field in type-II superconductors, which is equal to the thermodynamic critical field H_c in type-I materials[15]. The specimen is considered as massive if its minimal size significantly exceeds a so-called penetration depth λ, a width of the surface layer where B decays from its value on the external side of the specimen boundary B_{ext} to zero inside it. An example of magnetization curves of a type-I superconductor in the Meissner state is shown in Fig. 3.6.

3.1.4 Superconducting ring

The Meissner state is not the only equilibrium superconducting state. Other equilibrium states, called intermediate and mixed states, are discussed in the next chapter. In all equilibrium states superconductors are (1) diamagnetic (i.e., χ is negative) and (2) all these states occur only in singly connected ellipsoidal specimens. In multiply connected bodies the situation is drastically different. The difference arises due to a possibility of occurring the total current, impossible in the equilibrium states (see Sec. 1.2). Following the Faraday law, the total current is induced by the changing applied field to maintain the flux in the sample opening(s). Accordingly, the magnetic moment due to the

---

[15] A type-I superconductor can be viewed as a type-II one in which $H_{c1} = H_{c2} = H_c$.

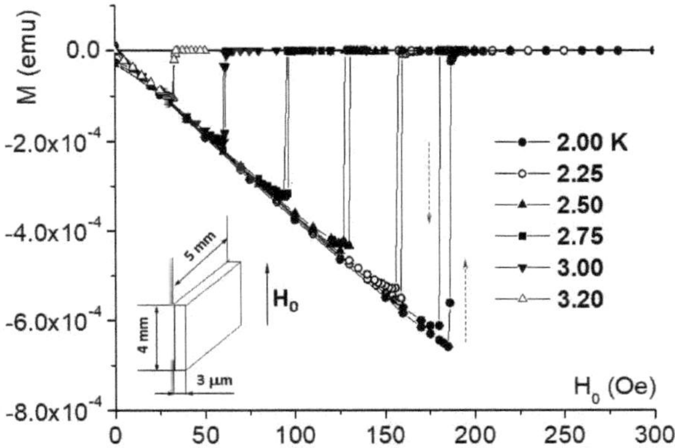

Figure 3.6   Magnetic moment of a type-I superconductor in the Meissner state. The specimen is an indium film with RRR $\approx$ 700; H_0 is the applied field. The data were taken at constant temperatures, as indicated. The insert shows the specimen/field configuration; in this case $\mathbf{H} = \mathbf{H_0}$. The specimen was cooled at zero field (arrow up for 2 K) and at $H_0 \approx 1.5 H_c$ (arrow down for 2 K). The hysteresis at the S/N transition is the supercooling effect caused by the positive S/N surface tension. After Kozhevnikov et al. [159]; reprinted with permission from Springer Nature.

total current is directed against the field *change*, i.e., it is negative when $\mathbf{H_0}$ is increased and positive when it is decreased.

The simplest and typical example of multiply connected bodies is a ring. The magnetic moment of a superconducting ring in the transverse field was measured by Shoenberg [108]; reported smoothed experimental data are shown in Fig. 3.7. The field distribution near the ring was measured by Smith and Wilhelm [104], and Shubnikov and Chotkevitch [109]. The ring magnetic properties were analyzed by Schoenberg [15]. Here we reproduce Schoenberg's analysis with minor modifications.

Let us consider a torus-like ring of a median radius R made of a wire of radius r, small compare to R. The wire material is lead (type-I superconductor). The ring is cooled down at zero field. The e.m.f. induced in the ring due to the changing H_0 is

$$\frac{\pi R^2}{c}\frac{dH_0}{dt} = L\frac{dJ_t}{dt}.$$
(3.1)

where L is the self-inductance of the ring and J_t is the total current encircling the ring.

Integrating this equation one obtains

$$cJ_t L = \pi R^2 (H_0 - H_i),$$
(3.2)

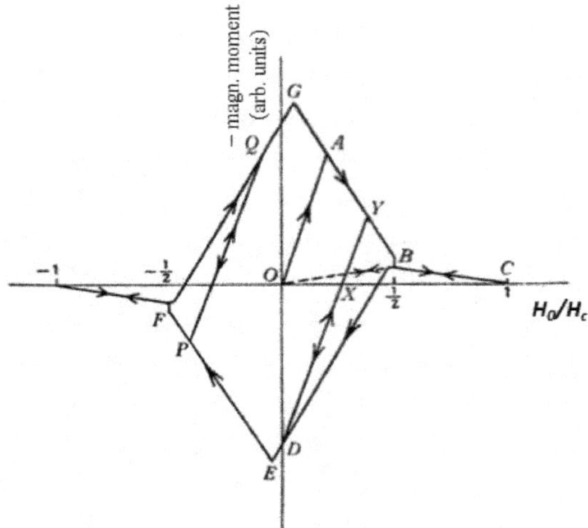

Figure 3.7 Experimental dependence of the magnetic moment of a superconducting ring on the field at a constant temperature. The field is applied transverse to the ring plane. Reprinted from [15] with permission from Cambridge University Press.

where H_i is the initial value of H_0 when J_t just appears. In the given case $H_i = 0$.

Current J_t creates a magnetic moment $M_t = \pi R^2 J_t / c$.

On the other hand, in the absence of J_t (e.g., if the ring is cut), the ring magnetic moment is equal to the moment of the superconducting wire M_w in the transverse field (in the given geometry, the effect of wire curvature is negligible). As we will see in the next chapter, when H_0 is less or equal to $H_c/2$ the wire is in the Meissner state. Hence, its magnetic moment is

$$M_w = -\frac{1}{4\pi}(2\pi^2 r^2 R)\,\frac{H_0}{(1-\eta)} = -\pi r^2 R H_0, \tag{3.3}$$

where the demagnetizing factor $\eta = 1/2$.

M_w is the moment due to induced microscopic currents, which we will denote as J_i. Providing the ring (wire) is superconducting, J_i and, therefore, M_w takes place regardless of the presence or absence of J_t. In our case the ring is closed and therefore (at least at the beginning) $J_t \neq 0$. Herewith on the OAB segment of the M-curve (Fig. 3.7) J_t runs along the outer rim of the ring. From Eq. 3.2, on the OA segment J_t is

$$J_t = \frac{\pi R^2}{cL} H_0. \tag{3.4}$$

The moment of this current M_t is negative (directed against the *increasing* $\mathbf{H_0}$ in the segments OA and AB) and significantly greater[16] than M_w, shown

---

[16]On the OA segment $M_t/M_w \sim (R/r)^2$.

by the dashed line OB. Important that on the OA segment both M_t and M_w moments have the same direction (against $\mathbf{H}_0$). This means that on the outer rim of the ring both J_t and J_i are parallel and increase with increasing H_0.

In p. A the sum of these currents reaches the critical value J_c, which according to Silsbee's hypothesis (see Sec. 5.2.4) means that the field at the rim reaches its critical value H_c. Therefore, the field at the outer rim in p. A is

$$\frac{H_0}{(1 - \eta)} + \frac{2J_t}{rc} = H_c. \tag{3.5}$$

Substituting $\eta = 1/2$ and J_t from Eq. 3.4, one can find the field H_0 in p. A

$$H_0 = \frac{H_c}{2}(1 + \pi R^2/Lrc^2)^{-1}. \tag{3.6}$$

This is the field starting from which the total current begins decreasing. The reason for this decrease is as follows. As soon as in some point of the rim the field reaches the critical value, the ring material in this point becomes normal and the flux quantum (see Sec. 5.2.5.) enters the ring. This means that lesser total current is needed to maintain the field within the ring and so this current is restored with a lower magnitude. After a small step up the process repeats again and again, resulting that the field at the rim stays equal to H_c in the AB segment. Coming from Eq. 3.5, the total current in this segment is

$$J_t = rc\left(\frac{H_c}{2} - H_0\right). \tag{3.7}$$

In, p. B the field within the ring becomes equal to the applied field and the current J_t ceases completely. As seen from Eq. 3.7, this happens at $H_0 = H_c/2$, i.e., at the field when the wire transitions to the intermediate state (see Sec. 4.3.2). This means that the wire material splits for S and N domains shaped as discs transverse to the wire [111, 110]. Hence, at $H_0 > H_c/2$ the wire can no longer support the total superconducting current in either way. At the same time the ring is still superconducting (meaning that J_i continues circulating in the S domains) until p. C where $H_0 = H_c$. Above this field the entire ring becomes normal.

On the way back, at decreasing the applied field, the process is reproduced in reversed order. The total current sets up again at p. B, but now it runs on the *inner* rim and keeps the field within the ring equal to $H_c/2$ (the initial field H_i at which the total current reappears). Important that in the inner rim the currents J_i and J_t run in opposite direction, as a result of which the arm BE is longer than EF. On the whole, the figure BEFGB is close to a parallelogram, which approaches a diamond shape when R/r increases.

For comparison with experimental data, Shoenberg used two approximate formulas for self-induction of the ring[17]. One of them is for a superficial current

---

[17]Analytical formulas for self-induction of the torus-like ring are always approximate [22]. An extended list of these formulas is available in [112].

and another one is for the current flowing through the whole cross-section of the wire. The self-inductance in the former differs from that in the latter for about 10%. A quantitative agreement was found using L for the superficial current.

To summarize, there are two kinds of superconducting currents. The first ones are microscopic currents, which are solely responsible for equilibrium properties, but exists both in the equilibrium and non-equilibrium states. The second one is the macroscopic or total current(s) appearing in multiply connected superconductors[18]. As mentioned above, the total current also includes the transport current in hybrid S/N circuits fed from an external source. The two kinds of superconducting currents resemble two kinds of currents in normal metals: the bound (non-dissipating) and total conduction (dissipating) currents. The principal difference is that in superconductors the currents of both kinds are dissipation free.

The total superconducting current mimics the current which could be expected in the hypothetical perfect conductors, but this in no way means that it is carried by "perfectly free" electrons.

3.2   THEORIES

In the following four sections, we consider the theories that have played a significant role in the interpretation of superconductivity. Those are the two-fluid model of Gorter and Casimir, the theory of F. and H. London, called the London theory, the theory of Ginzburg and Landau (the GL theory), and the theory of Bardeen, Cooper and Schrieffer (the BCS theory).

3.2.1   Two-fluid model of Gorter and Casimir

By the late 1920s, there were quite a few experimental results suggesting that thermodynamics can be applicable to superconductivity, and hence the latter may have only little to do, if any, with the perfect conductivity. This includes, in particular, a well established fact that the coexisting curve between S and N states in a magnetic field $H_c(T)$ has a parabolic shape for specimens of type-I superconductors with $\eta = 0$[19], i.e.

$$\frac{H_c}{H_{c0}} = 1 - \left(\frac{T}{T_c}\right)^2 , \qquad (3.8)$$

where H_{c0} is the critical field at zero temperature and T_c is the critical temperature at zero field.

---

[18]The term multiply connected superconductors covers an innumerable amount of non-ellipsoidal bodies with trapped magnetic flux. Moreover, insufficiently pure ellipsoidal bodies with pinned flux lines also represent the multiply connected objects.

[19]As is shown in the next chapter, the critical field of such specimens is the thermodynamic critical field.

Such a shape of the coexisting curve, as was shown by Keesom in 1924 (see footnote (8)) implies that the S/N transition in the field is a first order phase transition (the latent heat $Q \neq 0$), which becomes of the second order ($Q = 0$) at $T = T_c$ (see Sec. 4.3.2). One can see a close similarity between Keesom's formula (Eq. 4.11) and the Clausius-Clapeyron relation [93], and between the S/N and, e.g., liquid-gas transitions.

The discoveries of the jump in heat capacity and the Meissner effect put the end in argues about applicability of thermodynamics. Immediately after that Cornelis Gorter and Hendrik Casimir presented the first and highly seminal model of superconductivity, referred to as the two-fluid model [6].

The two-fluid model is a thermodynamic theory addressing properties of superconductors in zero field. Its key idea is that the conduction electrons of the superconducting material are divided for two interpenetrating fractions of different energy levels[20]. The energy difference between these levels is called *condensation energy* discussed in the next chapter. The fraction x with the higher (Fermi) energy represents "non-condensed" or "normal" electrons, correspondingly another fraction (1-x) represents "condensed" or "superconducting" electrons.

To characterize the condensed fraction, Gorter and Casimir postulated that entropy of the superconducting electrons is zero. Hence, electrons in this group are supposed to be completely ordered. As mentioned above, this postulate was based on the absence of thermoelectric effects in superconductors. The fractions are functions of temperature: $x = 0$ at $T = 0$, and $x = 1$ at $T = T_c$.

The free energy density[21] of conduction electrons $f(T)$ calculated relative to its value at zero temperature is taken as

$$f(T) = x^{0.5} f_n(T) + (1 - x) f_s(T), \tag{3.9}$$

where subscripts n and s designate the N and S fractions, respectively.

The free energy of the N fraction is chosen as $f_n = -aT^2/2$ to fit the linear dependence of the electron heat capacity in normal metals. The power 0.5 in the first term is picked so to fit the quadratic temperature dependence of the thermodynamic critical field $H_c(T)$ (Eq. 3.9), which reflects the cubic temperature dependence of the electron heat capacity in superconductors [113] (see Problem 3.1). The free energy of the S fraction is taken as $f_s = -b = const$, what follows from the zero entropy postulate. The coefficients a and b are parameters characterizing properties of the N and S fractions, respectively. In particular, a is the coefficient of proportionality in the temperature dependence of the heat capacity of normal electrons.

---

[20]This turned out very fruitful idea is used in all theories of superconductivity and superfluidity ever since.

[21]Since the system (specimen) is in zero field, there is no difference between the Helmholtz, Gibbs and total free energies. For the same reason the free energy does not depend on the specimen shape. Consequently, the concept of free energy density is applicable to samples of any shape [16].

At equilibrium $(\partial f/\partial x)_T = 0$. From that with the use of condition $x(T_c) = 1$ it follows that

$$x = \left(\frac{T}{T_c}\right)^4 \tag{3.10}$$

and

$$b = \frac{aT_c^2}{4}. \tag{3.11}$$

After substituting Eqs. (3.10) and (3.11) into Eq. (3.9) and using the same condition $x(T_c) = 1$, Eq. (3.9) takes the form

$$f = -a\frac{T_c^2}{4} - a\frac{T^4}{4T_c^2}. \tag{3.12}$$

Therefore, the electron specific entropy is

$$s = -\frac{df}{dT} = a\frac{T^3}{T_c^2}. \tag{3.13}$$

Note that, in spite of zero entropy of the superconducting electrons, the entropy of the S fraction is not zero due to the temperature dependence of $(1 - x)$, which is a relative number density of the superconducting electrons in the London theory.

In turn, the electron specific heat capacity is

$$c = T\frac{ds}{dT} = 3a\frac{T^3}{T_c^2}. \tag{3.14}$$

Referring to Kok [113], this justifies the choice of the power 0.5 in Eq. (3.9). Kok also showed that a can be expressed through parameters of the super-conducting state as (see Problem 3.1)

$$a = \frac{H_{c0}^2}{2\pi T_c^2}.$$

Accordingly, from Eq. (3.11) we find

$$b = \frac{H_{c0}^2}{8\pi}. \tag{3.15}$$

$H_c^2/8\pi$ is the condensation energy per unit volume (see Sec. 4.2). Therefore, b equals the condensation energy density at $T = 0$, which implies that the second term in Eq. (3.9) can be interpreted as a temperature dependent difference (gap) of the energies of electrons in the N and S fractions, similar to that which was later found in the BCS theory [38, 37].

All formulae of the model nicely fit experimental data. It is worth reminding that it was the two-fluid model (specifically Eq. (3.10)) that provided the success of the London theory and Eq. (3.10) itself is a demonstration of the predicting power of thermodynamics.

Overall, the broad list of successful formulae leaves no room to question the correctness of both main provisions of the two-fluid model: about inter-penetrating fluids and zero entropy. However, the specific physical content of these provisions remained to be explained.

3.2.2 London theory

The London theory was proposed by brothers Fritz and Heinz London in 1935 [114]; its final form is discussed in a book of F. London published in 1950 [35]. The London theory takes a significant part in the development of the superconductivity interpretation; suffice to say that it underlies the Ginzburg-Landau theory[36], Pippard's non-local theory [115] and the theory of Bardeen, Cooper and Schrieffer [37]. The London theory is reviewed in detail in [116]; here we partially reproduce this review.

The original goal of the theory, as was announced by the authors, was to modify an acceleration theory of Becker, Heller and Sauter [198] in such a way that it meets the Meissner effect. In [198] a superconductor is considered as a perfect conductor in which the superconducting electrons are totally free; i.e., their mean free pass is infinite.

Adopting this postulate, the Londons write

$$m\dot{\mathbf{v}} = e\mathbf{E}, \tag{3.16}$$

where m and e are the free electron mass and the electron charge, respectively, $\dot{\mathbf{v}}$ is the time derivative of its velocity; and $\mathbf{E}$ is the electric field acting on the electron.

Hence, the time derivative of the current density $\mathbf{j}(= n_s e \mathbf{v}$, where n_s is the number density of the superconducting electrons) is

$$\dot{\mathbf{j}} = \frac{c^2}{4\pi \lambda_L^2} \mathbf{E} \tag{3.17}$$

This is the first London equation, where λ_L is a material characteristic referred to as the London penetration depth. As postulated, λ_L is a function of temperature only. It is defined as[22]

$$\lambda_L = \left(\frac{mc^2}{4\pi n_s e^2} \right)^{1/2} \tag{3.18}$$

The second London equation is

$$\nabla \times \mathbf{j} = -\frac{c}{4\pi \lambda_L^2} \mathbf{H} \tag{3.19}$$

where $\mathbf{H}$ is the field intensity acting on the superconducting electrons.

---

[22]Note that if the superconducting electrons are combined in pairs, λ_L remains unchanged.

Eq. (3.19) is the core equation of the London theory, which, according to the authors, in superconductors replaces Ohm's law in the normal metals. In [114] it is derived from Eq. (3.17) taking curls from its both sides and using the Maxwell equation $c\nabla \times \mathbf{E} = -\partial \mathbf{B}/\partial T$ plus two more sets of postulates.

The first of them states that the magnetic permeability μ_m and the dielectric permittivity ε_e in the S phase are *unity*, i.e.,

$$\mu_m = \varepsilon_e = 1. \tag{3.20}$$

And the second set is

$$\mathbf{B}_\infty = \mathbf{j}_\infty = 0, \tag{3.21}$$

where subscript ∞ designates the quantities in the specimen interior or at the depth much greater than λ_L.

In other words, in the London theory (a) electromagnetic properties of the superconductors are identical to those of the vacuum (Eq. 3.20) and (b) there are neither fields, no currents in the specimen beyond the penetration depth (Eq. 3.21), implying that the field is expelled from the bulk of a superconductor in the MS.

The postulate Eq. (3.20), the identity of superconductivity to vacuum, was introduced as a simplification, which "may subsequently to be corrected" [114]. However, 15 years later F. London [35] keeps it in force and it has never been reconsidered afterward.

The first postulate in Eq. (3.21) ($\mathbf{B}_\infty = 0$) was needed to fit the Meissner effect, because in the acceleration theory the field inside specimen is frozen, i.e., it equal to an arbitrary field $\mathbf{H}_{in}$, which was there when the specimen lost its resistance. The second postulate ($\mathbf{j}_\infty = 0$) is justified by a never published theorem ascribed to Bloch (see [35]). On the other hand, $\mathbf{j}_\infty = 0$ directly stems from the symmetry of time reversal, provided that $\mathbf{j}_\infty$ is the average current density (see Fig. 1.1).

The postulates of Eqs. (3.20) and (3.21) automatically lead to

$$\mathbf{H}_\infty = \mathbf{v}_\infty = 0. \tag{3.22}$$

From Eq. (3.17) using the Maxwell equations $c\nabla \times \mathbf{H} = 4\pi \mathbf{j}$ and $\nabla \cdot \mathbf{B} = 0$ along with conditions $\mu_m = 1$ and $H_\infty = 0$ one obtains

$$\lambda_L^2 \nabla^2 \mathbf{H} = \mathbf{H}. \tag{3.23}$$

Analogically, with the use $\mathbf{j}_\infty = 0$ and taking into account that no external current is fed into the sample, one obtains

$$\lambda_L^2 \nabla^2 \mathbf{j} = \mathbf{j}. \tag{3.24}$$

Eqs. (3.23) and (3.24) imply that in the massive specimens (i.e., when the curvature of the specimen surface at distances of the order of λ_L can be neglected) the induction $\mathbf{B}(= \mathbf{H}$ in the theory) and the current density $\mathbf{j}$

exponentially decay with depth from their values at the specimen boundary to zero in its interior with the decay constant λ_L in both cases. Hence, taking into account the smallness of $\lambda_L (\sim 10^{-6}$ cm assuming one superconducting electron per atom), it looks as the theory meets the Meissner condition (zero B in the volume of the massive specimen) explaining it as a screening effect of eddy currents (*assumed* circumferential[23]) persistently running in a thin surface layer.

The reader has probably noticed that the theory represents a rather strange construction, the sole purpose of which is to fulfill the *manually* introduced Meissner condition $B = 0$[24]. Therefore, it was quite unexpectedly when it turned out that, being combined with the two-fluid model[25], results of the London theory appeared to be consistent with experimental data on the temperature dependence of the penetration depth [15].

Another form of the London equation (3.19) is

$$\mathbf{j} = -\frac{c}{4\pi\lambda_L^2}\mathbf{A}, \tag{3.25}$$

where $\mathbf{A}$ is the vector potential defined as $\mathbf{H} = \mathbf{B} = \nabla \times \mathbf{A}$ with the Coulomb gauge presumed (see Sec. 1.2.6).

Eq. (3.25) is that form of the London equation for which Pippard suggested a non-local extension of the London theory [115], which is also consistent with experiment [118, 119].

After all, Eq. (3.25) can be written (substituting $\mathbf{j} = en_s\mathbf{v}$ and λ_L from Eq. (3.18)) in a microscopic form as

$$m\mathbf{v} + \frac{e}{c}\mathbf{A} \equiv \widetilde{\mathbf{p}} = 0, \tag{3.26}$$

where $\widetilde{\mathbf{p}}$ is a generalized or canonical linear momentum of a single superconducting electron.

Eq. (3.26) is highly remarkable. F. London pointed out that the theory can be constructed starting from Eq. (3.26) as a postulate [35]. At the same time he did not reconsider the postulates (3.20) and (3.21). Note that, according to Eq. (3.25), $\mathbf{j}_\infty = 0$ implies that $\mathbf{A}_\infty = 0$ as well.

Here are some consequences of Eq. (3.26).

1. The London equation (3.19)/(3.25) takes an extremely simple form[26]: $\widetilde{\mathbf{p}} = 0$.

---

[23]Recall that the idea of the field induced circumferential surface current appeared well before the Meissner effect was discovered and the London theory proposed (see Sec. 3.1.1.). Nowadays the assumption that the induced current is the circumferential one is often taken for granted. However, this is one of the most controversial assumptions of the standard theories. One can clearly see this from the fact that the circumferential current represents a total current, which is impossible in equilibrium states (see Sec. 1.2.1).

[24]It should be noted that initially the theory was not well received. In particular, Shoenberg ignored it in the first edition of his book [15] (1938). See also footnote (13) in [116].

[25]As mentioned above, n_s/n_{s0} in the London theory corresponds to $(1-x)$ in the two-fluid model. Accordingly, $n_s/n_{s0} = 1 - (T/T_c)^4$, where n_{s0} is n_s at $T = 0$.

[26]Simplicity is the doubtless merit of any theory.

2. Taking de Broglie's wave length as $\lambda_{DB} = h/\widetilde{p}$, suggests that λ_{DB} of superconducting electrons is infinite, indicating that the superconducting body is a macroscopic quantum object.

3. Since the London theory is not restricted by any particular area of the phase diagram for the MS, Eq. (3.26) should be valid regardless on parameters of this state (e.g., temperature T and the applied field H_0). So F. London concluded that the S *phase*[27] possess a long-range order, characterized by the "rigid" (i.e., independent on the state parameters) generalized linear momentum $\widetilde{p}(= 0)$.

4. The rigidity concept was used by F. London to justify the reversibility of the effect of magnetization of superconductors by rotation, following from the acceleration theory of Becker et al. [198]. This so-called London moment was confirmed experimentally [64].

5. For multiply connected bodies the concept of rigidity led F. London to prediction of the magnetic flux quantization in superconductors [35]. A decade later it was confirmed experimentally [12, 13], albeit with a factor $1/2$ absent in the original prediction; this factor, as it was suggested by Onsager [120], is consistent with the pairing concept of the BCS theory [38]. The prediction of the flux quantization is, in any respect, the highest achievement of the London theory.

We see that the London theory has an impressive list of achievements. However, it is not hard to notice serious difficulties in this theory[28]. Let us look closer at some of them.

For convenience, we rewrite here Eq. (1.30) for the magnetic moment M of a specimen in the Meissner state ($\mu_m = 0$) and Eq. (2.36) for the magnetic energy of the cylindrical specimen in the same state

$$\mathbf{M} = -\frac{V}{4\pi} \cdot \frac{\mathbf{H}_0}{(1 - \eta)} \tag{1.30}$$

$$E_m \equiv -\int \mathbf{M} \cdot d\mathbf{H}_0 = \frac{V H_0^2}{8\pi} = \sum_s \frac{m v_i^2}{2} = E_k, \tag{2.36}$$

where $m v_i^2/2$ and E_k are kinetic energies of the field-induced motion of a single and all superconducting electrons, respectively; $\mathbf{v}_i$ is the induced velocity. The summation is taken over all superconducting electrons in the specimen. Eq. (1.30) follows from definition of the magnetic moment, and Eq. (2.36) reflects the law of energy conservation for diamagnetic specimens of the cylindrical geometry (see Sec. 2.3.2).

---

[27]The S phase is defined as a phase of a superconducting material in which $n_s \neq 0$. The S phase can occupy the entire sample volume (as it takes place in the MS) or a part of it.

[28]From the very beginning (see [114]) the Londons well understood the weakness of the theoretical background of Eqs. (3.17) and (3.19). In [35] F. London does not derive them stating instead that the validity of these equations follows from the experimental confirmation of the consequences which they imply.

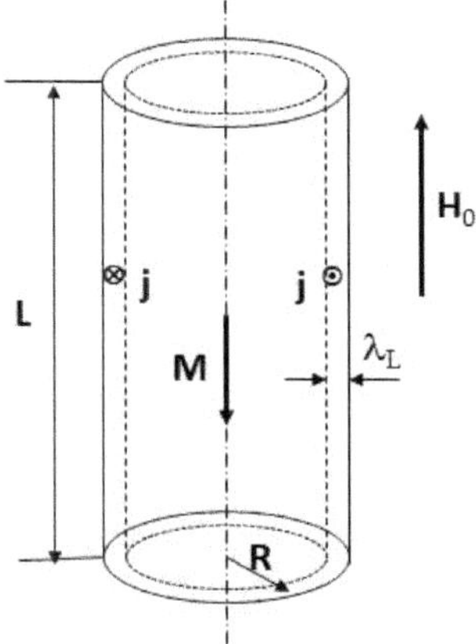

Figure 3.8 A cylindrical specimen in the London theory. $L \gg R \gg \lambda_L$, the demagnetizing factor $\eta = 0$. The London penetration depth λ_L is a constant of an exponentially decaying induction; it is also an effective width of a surface layer with the field induced screening current. The induction **B** and the current density **j** in this layer are equal to $\mathbf{B}_0 = \mathbf{H}_0$ and $c\mathbf{n} \times \mathbf{H}_0/4\pi\lambda_L$ with **n** being a unit vector normal to the surface, respectively; beyond this layer, **B**, **H**, **A** and **j** are zero. **M** is the specimen magnetic moment caused by the surface current. $\mathbf{H}_0$ is the applied field.

A schematic of the cylindrical specimen in the London theory is shown in Fig. 3.8[29]. An "active" part of this specimen, where the field (both **B** and **H**) and the current **j** are not zero, is a surface layer with an effective width[30] $\lambda_L \ll R$. Behind this layer all magnetic characteristics are zero, implying that the specimen interior, practically whole its volume, is totally inert. If so, there should be no difference if the interior is in the S or in the N state, or there is no interior at all.

Thus, in the London theory magnetic properties of solid (continuous) and hollow specimens are identical (see also Sec. 5.2.11). However, as is well known [1, 121], properties of the solid and hollow magnetized bodies are different. For superconductors it was clearly demonstrated by Meissner and Ochsenfeld.

More specifically, in the London theory a long solid cylinder in the longitudinal field $\mathbf{H}_0$ is identical to an empty solenoid of the same shape and in

---

[29]In [35] and [15] solutions of the London equations for the field, current and the magnetic moment are available for cylinders, plates and spheres. The calculated magnetic moment of the cylinder and sphere with radius $R \gg \lambda_L$ equals the moment of these figures with zero induction and the radius $R - \lambda_L$; for the plate of thickness $d \gg \lambda_L$ the moment equals that of the plate with $B = 0$ having the thickness $d - 2\lambda_L$. The surface layer with the effective width λ_L represents an excluded volume, which is discussed in Sec. 5.2.9

[30]The effective width of the penetration layer is defined as $\lambda_{eff} \equiv H_0^{-1} \int_0^\infty B(z)dz$, where z is the depth from the surface and $B(z)$ is the depth's profile of the induction. Inside this layer $B = B_{eff} \equiv (\lambda_{eff})^{-1} \int_0^\infty B(z)dz$ and $j = j_{eff} \equiv (\lambda_{eff})^{-1} \int_0^\infty j(z)dz$, where $j(z)$ is the depth's profile of the current. In the cylindrical sample of the London theory $\lambda_{eff} = \lambda_L$, $B_{eff} = H_0$ and $j_{eff} = g/\lambda_L = cH_0/4\pi\lambda_L$, where g is the linear current density.

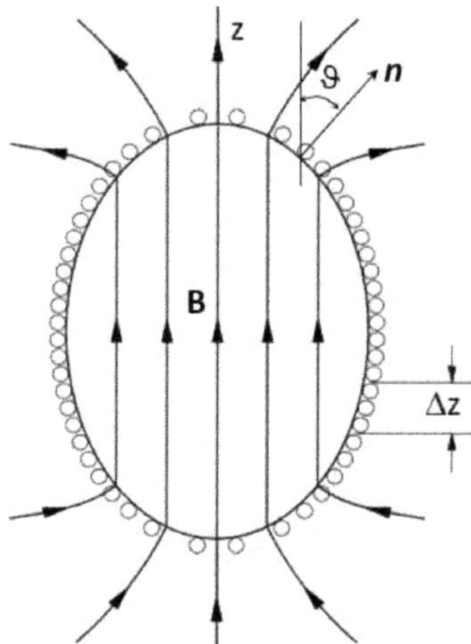

Figure 3.9 A current shell having the shape of an ellipsoid of revolution (spheroid) with respect to the vertical axis z; ϑ is the angle between z and the unit vector **n** normal to the shell. If the current linear density g is proportional to $\sin\vartheta$ or, equivalently, there are equal currents in each axial interval Δz, the magnetic field inside this shell $\mathbf{B}(=\mathbf{H}$ since $\mu_m = 1)$ is uniform and parallel to z.

the same field with the current per unit length $\mathbf{g} = c\mathbf{n} \times \mathbf{H}_0/4\pi$[31], where **n** is the unit vector normal to the surface and directed outward; the moment **M** produced by this current is the same as that in Eq. (1.30a) with $\eta = 0$ (see problems 3.2 and 3.3). The field inside the solenoid due to this current equals $-\mathbf{H}_0$, it compensates the field $\mathbf{H}_0$ resulting that the field there (both **H** and **B** since the solenoid is empty and therefore $\mu_m = 1$) is zero. Hence, the *solenoid's interior* is screened from the applied field, or the field is "expelled" from the solenoid when the current is turned on.

The same statement about the screening current and the expelled field holds for a spheroidal current shell (the solenoid is a particular case of such a shell) shown in Fig. 3.9.

On the contrary, inside specimens in the MS the induction $\mathbf{B} = 0$, but the intensity **H** is not (see Sec. 1.2.5 and Fig. 3.10). In a cylinder in the parallel field, like that in Fig. 3.8, $\mathbf{H} = \mathbf{H}_0$. However, sometimes one can read that the field due to the induced circumferential surface current compensates the field **H** resulting in zero field (both B and H as stated in the London theory) inside the specimen[32].

---

[31]This current is calculated from the boundary condition for B_t (Eq. 1.24), which stems from the *continuity* of H_t (Eq. (1.18))

[32]A statement of such kind (appeared originally in the acceleration theory of Becker et al. [198]) can be found in some textbooks, however it is incorrect because it contradicts to the continuity of H_t (Eq. (1.18)) which is used to calculate the surface current (Eq. (1.25)).

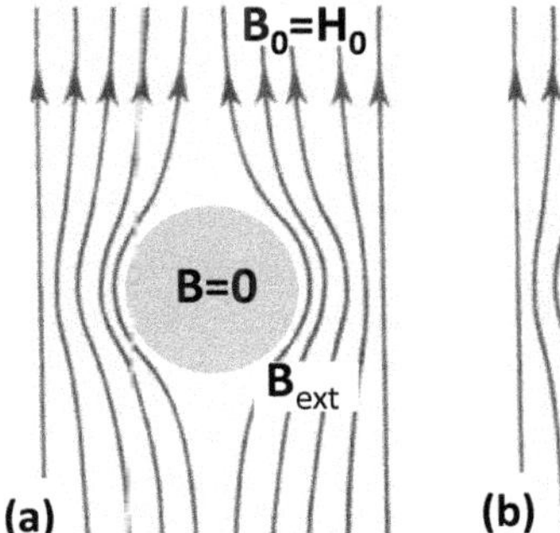
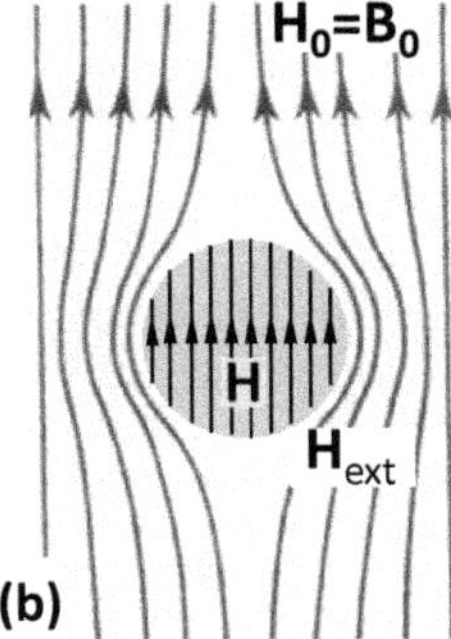

Figure 3.10   Induction **B** (a) and intensity **H** (b) inside and outside of a long cylindrical specimen in the Meissner state when $\mathbf{H}_0$ is perpendicular to its longitudinal axis; in this case $\eta = 1/2$. Inside the specimen $\mathbf{B} = 0$ and $\mathbf{H} = 2\mathbf{H}_0$; outside it $\mathbf{H} = \mathbf{B}$: near the specimen the field $\mathbf{H}_{ext} = \mathbf{B}_{ext}$ is tangential to the surface, and far away it is equal to the applied field $\mathbf{H}_0 = \mathbf{B}_0$. Shadowed circle is a cross-section of the specimen oriented perpendicular to the page. (Reprinted from [169] with permission from Elsevier.)

Now, taking into account what was said about the spheroidal current shell, can one claim that the assumption of the induced circumferential current provides a consistent picture of the MS? The answer is *no* already because the MS is observed in specimens of any ellipsoidal shape, but not only in the ellipsoids of revolution. For example, it can be the cylinder in the transverse field shown in Fig. 3.10 [15] or a film in the parallel field, like the film which magnetization curves are shown in Fig. 3.6. Another reason of this negative answer is that the field inside the spheroidal specimen, produced by the circumferential current required to obtain the correct magnetic moment, equals $-\mathbf{H}_0$, while the field which is supposed to be compensated is $\mathbf{H}_0/(1 - \eta)$ (see problem 3.5). On top of that the circumferential current in the MS is just never possible since it would violate the symmetry of time reversal, which is mandatory for any equilibrium state.

Another difficulty of the London theory can be seen from the following. A specimen with $\eta = 0$ can be an infinite plate in the parallel field, like the film in Fig. 3.6. According to the London theory, the applied field $\mathbf{H}_0$ induces a circumferential screening current persistently running along the film surface within a thin layer where the field is not zero; a transverse (perpendicular to $\mathbf{H}_0$) cross-section of this current represents a rectangle with an infinite length/width ratio ($\sim$ 5 mm/3 μm). But, how can electrons (supposed free) run along the straight path if they are constantly pushed sideways by the magnetic part of the Lorentz force? How can they make a U-turn at the ends keeping the constant speed, without canceling the law of inertia and following from it Newton's acceleration equation Eq. (3.16)? The London theory does not answer these questions.

Even clearer picture is as follows. Consider the cylindrical specimen in Fig. 3.8. Coming from the boundary conditions for $\mathbf{B}_t$ (Eq. (1.24)) and using the known formula for $\mathbf{j}$, one can calculate average kinetic energy ϵ_k of the field induced motion of one superconducting electron. This energy is $\epsilon_k = H_0^2/8\pi n_s$ (see problem 3.7). From that one can find kinetic energy E_k of the induced motion of all superconducting electrons in the effective penetration layer, which create the correct magnetic moment $\mathbf{M}$. This energy is

$$E_k = \frac{H_0^2 V}{8\pi} \cdot \frac{2\lambda_L}{R}, \tag{3.27}$$

Note that this formula directly follows from the Londons' assumption of the circumferential screening current, according to which superconducting electrons are active only in the penetration layer.

According to the law of energy conservation, the magnetic energy of this specimen $E_m (= H_0^2 V/8\pi)$ equals the field induced kinetic energy of electrons E_k (see Eq. (2.36)). However, in the London theory, as seen from Eq. (3.27), $E_k \ll E_m$ which clearly contradicts the law[33].

The conflict of the London theory with the energy conservation was noticed long ago by Shoenberg [15]. He paid attention to the fact that in a spherical specimen in the MS the assumption of circumferential current leads to an appearance of an e.m.f. between points at the pole and the equator. If so, it would be possible to continuously draw energy from the static magnetic field into a resistive load connecting these points, which clearly contradicts the law of conservation of energy.

Any *one* of the listed inconsistencies (this list can be continued [116]) is sufficient to cast doubt on the London theory. However, still one more very strange thing is that the theory does not mention the necessary condition of the existence of dissipation-free current: quantization of the angular momentum of its carriers.

Summarizing, in the London theory an attempt was made to apply the concept of perfect conductivity to describe the magnetic properties (primarily the Meissner effect) of superconductors. To reach this goal the authors were forced to sacrifice the fundamental physics laws (the conservation of energy, the symmetry of time reversal, the law of inertia, the Faraday law, etc), the principles of electrodynamics (both tangential and normal components of the magnetic field $\mathbf{B}=\mathbf{H}$ at the specimen boundary are continuous in the theory) and, finally, the common sense (postulate of equivalence of superconductivity to vacuum). Nevertheless, as it has happened in the history of physics, the London theory, despite its complete inadequacy, proved to be useful. The latter is clearly seen from the listed above achievements of the theory.

---

[33]Note, that if each superconducting electron of the specimen acquires kinetic energy ϵ_k, then $E_k = n_s \epsilon_k V = V H_0/8\pi$: the law is met!

3.2.3 Ginzburg-Landau (GL) theory

One of disadvantages of the London theory is its failure to explain an excess S/N interphase energy (positive surface tension) necessary for a description of the intermediate state (see Sec. 4.3.2). To solve this problem was a major goal of the Ginzburg-Landau theory developed in 1950 [36]. Apart from the original paper, the theory is described in nearly all textbooks on superconductivity, therefore here we restrict ourselves to a brief overview of its main achievements and downsides.

The GL theory is based on Landau's thermodynamic theory of the second order phase transitions, in which an order parameter, characterizing an emerging new ("ordered") phase is considered as a small perturbation of the system free energy [43]. In the GL theory the order parameter Ψ represents the superconducting component of the conduction electrons. Ψ is taken in a complex form as $\Psi = \Psi_0 \exp(i\varphi)$, where the amplitude $\Psi_0 = |\Psi|$ is determined by the density of superconducting electrons n_s ($\Psi_0^2 = n_s$), whereas the phase gradient $\nabla\varphi$ is proportional to the density of superconducting current $\mathbf{j}_s$, as it takes place in quantum mechanics. The authors note that the order parameter treated in this way can be considered as an average wave function of superconducting electrons. The complex form of the order parameter of the GL theory (i.e., the introduction of phase φ) was used by Josephson to justify predicted him effect [123] (see Sec. 5.2.7). The contribution of superconducting electrons in the given theory is small by definition (i.e., $n_s \ll n$, where n is the total density of conduction electrons). This means that a range of the theory applicability is restricted by the close vicinity to the critical temperature T_c [124].

The most important achievement of the GL theory is the demonstration that properties of superconductors are controlled by two microscopic parameters of the dimension of length. In the GL theory these parameters are the London penetration depth[34] and a so-called Ginzburg-Landau coherence length ξ. Like in the London theory, λ_L represents a length scale over which the induction decays from its value in the N phase to zero in the interior of the S phase. In turn, ξ is the length scale on which $|\Psi|$ changes from zero in the N phase to a constant value $|\Psi_\infty|$ in the S phase. The ratio λ_L/ξ is the Ginzburg-Landau parameter κ. Near the critical temperature (at $T/T_c \to 1$) κ is a fixed value different for different materials.

In superconductors with $\kappa < 1/\sqrt{2}$ (or $\xi - \lambda_L\sqrt{2} > 0$) the S/N surface tension is positive; such materials are referred to as type-I superconductors. And vice versa: if $\kappa > 1/\sqrt{2}$, the surface tension is negative; materials with such κ are called type-II superconductors. Properties of type-I and type-II superconductors are discussed in the next chapters. Here we note that the factor $\sqrt{2}$ in the GL theory is due to the chosen dimensionless units in which

---

[34]In contrast to the London theory, where λ_L depends only on temperature, in the GL theory λ_L depends also on the field.

the critical field $H'_c = 1/\sqrt{2}$; if these units are taken such that $H'_c = 1$, then $\sqrt{2}$ disappears [125].

Type-II superconductors were discovered by Shubnikov with co-workers in 1937 [126]. Twenty years later, the application of the GL theory with the assumption of negative S/N surface tension allowed Abrikosov [127] to explain magnetization curves reported in [126].

On the downside, the GL theory by default accepts London's paradigm of the field expulsion, i.e., identity of **H** and **B** and, as a consequence, the circumferential screening current. This is the main reason of inadequacy of the GL theory[35].

3.2.4 Bardeen-Cooper-Schrieffer (BCS) theory

In 1956, just within a year after the first acquaintance with superconductivity [131], Leon Cooper published a theoretical paper [11], where he showed that in the presence of an attraction between electrons with energy close to the Fermi level (no matter how weak this attraction is), the Fermi liquid of conduction electrons becomes unstable with respect to formation of bound pairs, called Cooper pairs. The pairs have zero linear momentum with respect to their center of mass and compensated spins. Accordingly, Cooper pairs represent effectively spinless charge carriers.

A theory based on Cooper's idea of the bound states of conduction electrons and the associated energy gap in the electron spectrum [132] was developed by John Bardeen, Leon Cooper and Robert Schrieffer in 1957; the theory is known as the BCS theory.

The electron pairing implies a non-local character of the pair response on the magnetic field, similar as it was proposed by Pippard in his non-local theory [115]. In the latter, the local vector potential of London's equation Eq. (3.25) is replaced by the potential averaged over a region with dimension of ξ_0, called the Pippard coherence length. Similar notion of the coherence length is used in the BCS theory, where ξ_0 characterizes the dimension of Cooper pairs. Remarkably that ξ_0 calculated in the BCS theory nearly quantitatively matches the empirical value of Pippard's coherence length estimated from measurements of the penetration depth. This was one of triumphs of the BCS theory. Important to note that in pure metals $\xi_0 \sim 10^{-4}$ cm [11], which means a strong overlap of pairs; however, this fact does not affect the resistance-less current carried by the pairs.

---

[35]In the original paper of 1950 [36] the authors suggest to verify the theory by measuring the field dependence of the penetration depth, arguing that λ should depend on H_0. Independently, in the same year the field dependence of λ was measured by Pippard [128], who found a small and irregular field dependence (see Sec. 5.2.9). However, a quarter of a century later, Pippard noted that the problem was still open [129]. Ultimately, this problem was solved only recently owing to the novel possibilities provided by the low-energy μSR spectroscopy [130, 116]: the penetration depth does not depend on the field.

In the BCS theory the electron-electron attraction needed for stable pairing arises due to electron-phonon interaction, which leads to so-called isotope effect discovered earlier[36] [133, 134]. According to the latter, $\sqrt{M_a}T_c = const$, with M_a the atomic mass of an isotope. This formula is consistent with the BCS theory. Other aspects of the theory are discussed in Schrieffer's monograph [38] and in many textbooks.

Soon after publication of the BCS theory, its key concept, the Cooper pairing, was confirmed in already mentioned experiments of Deaver and Fairbank [12], and of Doll and Näbauer [13]. Both these experiments were targeted to verify the flux quantization predicted by F. London [35]. According to F. London, the flux Φ in multiply connected superconducting bodies is a multiple of $\Phi_0 = hc/e$, where h is the Planck constant. In the experiments it was found that the flux is indeed quantized, but the charge in denominator of Φ_0 equals $2e$, which confirms electron pairing [120].

To summarize, by incorporating electron pairing, the BCS theory made a key contribution to the interpretation of superconductivity. The pairing concept made it possible to abandon the unphysical notion of the "superconducting electrons". Cooper pairs represent a composite part of conduction electrons not present in "normal" metals. Clearly, that it is Cooper pairs that are responsible for unusual properties of superconductors.

Unfortunately, similar to the GL theory, the BCS adopts the Londons' paradigm of the field expulsion[37]. In result, in the BCS current induced in the Meissner state is the same as the total current in the ring, zero entropy (treated similar as in the two-fluid model), zero induction (postulated as such in the London theory) and zero resistivity (accepted by default) remained a mystery.

Completing this section about the theories of superconductivity, we would like to remind the reader that there is no and never has been a direct path in understanding phenomena of nature. Each of the considered theories has made an important contribution to solving the puzzle of superconductivity and together they provided the basis for finding its solution. At the same time, an adequate theoretical model must include and explain achievements of these theories. A model of such kind is discussed in Ch. 5.

3.3   PROBLEMS

3.1. Show that if the heat capacity of superconducting electrons $C_s \sim T^3$, the thermodynamic critical field H_c is a parabolic function of temperature T, as in Eq. (3.8) [Kok, 1934 [113]].

---

[36]Matthias challenged the generality of the isotope effect [135].

[37]Citing the authors "Our theory of the diamagnetic aspects follows along the general lines suggested by London and by Pippard" [37].

Solution. At zero field

$$\frac{C_s}{V} = BT^3$$

and

$$\frac{C_n}{V} = \alpha T + \beta T^3,$$

where in the right side the first term is the heat capacity of free (normal) electrons and the second term is the heat capacity of the lattice.

Hence,

$$\frac{C_s - C_n}{V} = (B - \beta)T^3 - \alpha T = \frac{T}{V}\frac{\partial(S_s - S_n)}{\partial T}$$

Taking into account the Third law, after integration on temperature from $T = 0$ to T, we get

$$\frac{S_s - S_n}{V} = \frac{1}{3}(B - \beta)T^3 - \alpha T = \alpha\frac{T^3}{T_c^2} - \alpha T,$$

since at $T = T_c$, $S_s = S_n$ and therefore $\alpha = (B - \beta)T_c^2/3$.

Now, from the definition of the thermodynamic critical field (Eq. (4.2)),

$$\frac{S_n - S_s}{V} = -\frac{\partial}{\partial T}\left(\frac{F_n - F_s}{V}\right) = -\frac{1}{8\pi}\frac{dH_c^2}{dT}.$$

So,

$$\frac{dH_c^2}{dT} = 8\pi\alpha\left(\frac{T^3}{T_c^2} - T\right).$$

After one more integration on temperature, we obtain

$$H_c^2 = 4\pi\alpha\left(\frac{T^4}{2T_c^2} - T^2\right) + H_{c0}^2.$$

At $T = T_c$, the critical field $H_c = 0$, so

$$\alpha = \frac{H_{c0}^2}{2\pi T_c^2}.$$

Substituting α into the previous equation, we obtain

$$H_c^2 = H_{c0}^2\left[1 - \frac{T^2}{T_c^2}\right]^2.$$

This is Eq. (3.8).

3.2. A long cylindrical solenoid consists of a single layer of tightly wound turns of thin wire. The solenoid length and radius are L and R, respectively; the wire diameter is D, and number of turns is N. Assume $L \gg R \gg D = L/N$. Solenoid is in free space and a uniform magnetic field H_0 is applied parallel to its longitudinal axis.

(a) What current J_0 should be set in solenoid to zero the induction B inside it? (b) Find magnitude of the linear density of this current g. (c) What is magnitude of the field strength H inside the solenoid? (d) What are B and H outside the solenoid? (e) What is magnetic moment of the unit length of the solenoid M/L?

Solution.
(a) The induction inside the solenoid $B_i = H_i$ since the solenoid is empty. At zero applied field, from the Ampére law, the field H_i is

$$H_i = \frac{4\pi}{c} n J_0,$$

where $n = N/L$.

When the applied field H_0 is turned on, the induction inside the solenoid becomes $B = B_0 + B_i = 0$, where $B_0 = H_0$. So the current J_0 needed to zero B is

$$J_0 = \frac{c}{4\pi} \frac{H_0}{n}.$$

(b)

$$g = \frac{J_0 N}{L} = J_0 n = \frac{c}{4\pi} H_0.$$

(c) $H = B = 0$.
(d) Solenoid is long (infinite), so everywhere outside it $B = H = H_0$.
(e)

$$\frac{M}{L} = -\frac{1}{L} \cdot N \frac{J_0 \pi R^2}{c} = -n J_0 \frac{\pi R^2}{c} = -\frac{1}{L} \frac{V}{4\pi} H_0,$$

where $V = \pi R^2 L$ is the solenoid volume. Sign $(-)$ means that direction of J_0 is such that the field H_i and therefore the moment M are anti-parallel to H_0.

3.3. A superconducting cylinder has the same dimensions (L and R) as those of the solenoid in the previous problem. The cylinder is in the Meissner state in the field H_0 parallel to its longitudinal axis; it is in vacuum.

(a) Find linear density of the surface current $\mathbf{g}_b$. (b) What are magnitudes of B, H and M for this specimen? Are magnetic properties of the superconducting cylinder are identical to those of the solenoid of the previous problem with the linear density of the surface current $g = cH_0/4\pi$? If not, then which properties are the same and which ones are different?

Answer: (a) $\mathbf{g}_b = -c\mathbf{H}_0 \times \mathbf{n}/4\pi$; (b) inside $\mathbf{B} = 0$ and $\mathbf{H} = \mathbf{H}_0$, outside $\mathbf{B} = \mathbf{H} = \mathbf{H}_0$; $\mathbf{M} = -V\mathbf{H}_0/4\pi$. So, the field intensity $\mathbf{H}$ inside the superconducting cylinder is different from that inside the solenoid ($= 0$). The rest of these properties are formally the same, although g in solenoid is the linear density of the transport current, whereas g_b in the superconductor is the linear density of the bound currents (the ones induced in Cooper pairs). Accordingly, at the boundary of solenoid there is a jump in H_t (Eq. (1.21)), whereas at the boundary of the superconductor the jump is in B_t (Eq. (1.25)), while H_t is continuous.

3.4. Find the linear density of the surface current g_b in the spherical superconducting specimen if the applied field is H_0 and the specimen is in the Meissner state.

Solution

Coming from the boundary condition for the tangential component of induction (1.24) and using H_{ext} calculated in the problem 1.10, we write

$$g_b = \frac{c}{4\pi(1-\eta)} H_0 \sin\theta = \frac{3c}{8\pi} H_0 \sin\theta.$$

Direction of $\mathbf{g}_b$ is found from Eq. (1.10). It is clockwise, as seen from the tip of $\mathbf{H}_0$.

3.5. A nonmagnetic ball of radius R is wound by a thin wire so that the turns are parallel to each other as shown in Fig. 3.11. Assume that each turn is a closed and electrically isolated circular loop. Now imagine that there is a current in each loop $J_\theta = gD$, where the wire diameter $D \ll R$ and g is linear density of the surface current numerically equal to the bound current density g_b obtained in the previous problem, and the current runs clockwise if seen from the top pole P. So, the loops make a shell with currents along the parallels. Show that the field inside the shell is uniform and find the field value.

Hints: (a) Calculate the field at the shell center (p. O) and at the poles (pp. P and P'). (b) The field at a distance z above the center of a circular

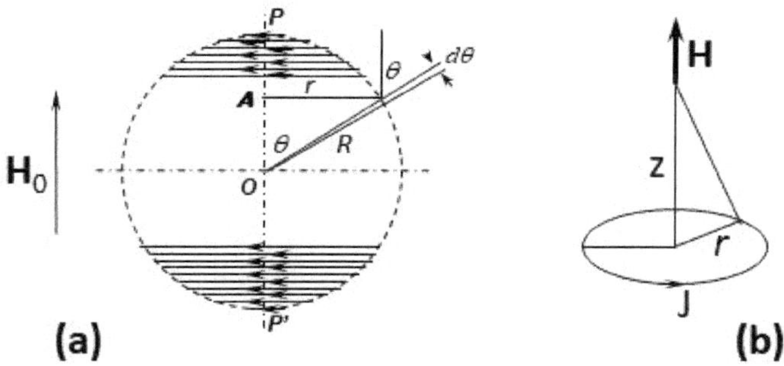

Figure 3.11 (a) Schematics of the current shell for problems 3.5 and 3.6. Horizontal lines depict the circular loops (not all loops are shown) lying in horizontal (perpendicular to the page) plane with current flowing clockwise, as shown by arrows. H_0 is the applied field. (b) Schematic of a magnetic field H at a distance z from a center of a circular loop of radius r carrying a steady current J.

loop of radius r with a current J (Fig. 3.11b) is

$$H = \frac{2\pi r^2 J}{c(r^2 + z^2)^{3/2}}$$

Solution.
A. Field at p. O.
Consider an arbitrary loop centered at p. A. Current dJ in this loop is

$$dJ = J_\theta = gRd\theta = \frac{3cRH_0}{8\pi} \sin\theta d\theta$$

So, the field dH_O due to this current in p. O is

$$dH_O = \frac{2\pi r^2 dJ}{cR^3} = \frac{2\pi(R^2 \sin^2\theta)dJ}{cR^3} = \frac{3H_o}{4} \sin^3\theta d\theta$$

And the total field H_O at p. O is

$$H_O = \frac{3H_o}{4} \int_0^\pi \sin^3\theta d\theta = H_0.$$

B. Field at p. P.
We write for the field dH at p. P due to current in the same single loop
centered at p. A

$$dH \quad = \quad \frac{2\pi(R^2 \sin^2\theta^2)dJ}{c(R^2 \sin^2\theta + R^2(1 - \cos\theta)^2)^{3/2}} \quad = \quad \frac{3H_o}{4} \cos^3(\frac{\theta}{2})d\theta$$

Therefore, for the field at p. P H_P we obtain

$$H_P = \frac{3H_o}{4} \int_0^\pi \cos^3(\frac{\theta}{2})d\theta = \frac{3H_o}{2} \int_0^{\pi/2} \cos^3\varphi d\varphi = H_0.$$

C. Field in p. P'.
Due to symmetry, the field in the lower pole P' is the same as that in the
upper pole P. Hence, $H_{P'} = H_0$.

Taking into account the direction of the currents, we see that the field in the
center and at the poles is the same and equals to $-\mathbf{H}_0$, where $\mathbf{H}_0$ is the applied
field causing the surface bound current g_b in the spherical superconductor
(problem 3.4). We leave it to the reader to show that the field is the same
$-\mathbf{H}_0$ in any other point inside the shell (for this purpose it is suffice to calculate
the field at the equatorial points using the boundary condition (1.20)). Note
that the field $\mathbf{H}$ inside a solid sphere is $\mathbf{H} = 3\mathbf{H}_0/2$, i.e., it is different from
the field inside the shell.

3.6. (a) What is magnetic moment of the current shell considered in the
previous problem? (b) If the shell is placed in vertically directed magnetic
field H_0 (Fig. 3.11), what are the fields H and B inside and outside the shell?

Solution (a) Magnitude of the magnetic moment dM due to the current in the single loop of the previous problem is

$$dM = \frac{\pi r^2 dJ}{c} = \frac{\pi R^2 \sin^2 \theta}{c} \cdot \frac{3cH_0 R}{8\pi} \sin \theta d\theta$$

Hence, the magnetic moment of the shell is

$$M = \frac{3H_0 R^3}{8} \int_0^\pi \sin^3 \theta d\theta = \frac{H_0 R^3}{2} = \frac{3}{2}\frac{V H_0}{4\pi} = \frac{V H_0}{4\pi(1 - 1/3)},$$

where $V = 4\pi R^3/3$ is volume inside the shell.

Note that M of the shell is the same as that of superconducting sphere ($\eta = 1/3$) in the Meissner state, as it should be because all bound currents in the bulk are compensated.

(b) Taking into account the direction of the field produced by the current in the loops (downward in Fig. 3.11) and the fact that μ inside the shell is unity, the fields inside are

$$\mathbf{H} = \mathbf{H}_0 + \mathbf{H}_{shell} = 0$$

$$\mathbf{B} = \mu \mathbf{H} = 0$$

The outer field near the shell is determined by its magnetic moment. Since the latter is the same as in the solid superconducting sphere, the external field near the shell is

$$B_{ext} = H_{ext} = \frac{H_0}{(1 - \eta)} \sin \theta = \frac{3H_0}{2} \sin \theta.$$

And the field far away from the shell is

$$B = H = H_0.$$

3.7. Calculate kinetic energy of a single superconducting electron near the specimen surface in the London theory. Consider the specimen of cylindrical geometry in the Meissner state (Fig. 3.8).

Solution

$$v = \frac{j}{n_s e} = \frac{g}{n_s e \lambda_L} = \frac{H_0 c}{4\pi n e \lambda_L},$$

where g is taken from (1.24). Hence,

$$\frac{mv^2}{2} = \frac{H^2 c^2 m}{2(4\pi)^2 n_s^2 e^2 \lambda_L^2} = \frac{H_0^2}{8\pi n_s}\frac{mc^2}{4\pi e^2}\frac{1}{\lambda_L^2} = \frac{H_0^2}{8\pi n_s}.$$

CHAPTER 4

SUPERCONDUCTIVITY. THERMODYNAMIC PROPERTIES

The existence and widespread occurrence of superconductivity is due to the thermodynamic advantage of the S state: the free energy of a superconducting specimen is less than that in the N state. For this reason thermodynamics takes an important part in all aspects of superconductivity. This chapter is devoted to thermodynamics of superconductors.

Recall that thermodynamics deals with the macroscopic properties of statistical systems (i.e., systems consisting of a statistically significant number of identical constituents) in equilibrium states. Those are the states, which properties (called thermodynamic) are completely determined by the state parameters or thermodynamic variables, such as temperature, volume, etc., regardless of the path by which the system is brought to the present state. Entropy of a system in equilibrium state is constant and maximum of its accessible values; this is the Second law of thermodynamics. Consequently, properties of an equilibrium system are reversible and time-independent, meaning that the symmetry of time reversal is fulfilled automatically.

Other laws of thermodynamics are the First law (energy conservation) and the Third law, also referred to as Nernst's theorem. The latter states that entropy of a system tends to zero when its temperature goes to zero and vice versa due to the reversibility of thermodynamic properties.

By definition, entropy is a measure of disorder in statistical systems. Therefore, the Third law is often interpreted as the impossibility of achieving zero temperature, since zero disorder (complete order) in such systems (like, e.g., crystalline solids) is unattainable. However, in some magnetics, specifically in diamagnetics, the ordering of microscopic magnetic moments is absolute since the number of accessible states for these moments is strictly 1. Therefore, magnetic component of entropy S_m in diamagnetics is zero (see problem

DOI: 10.1201/9781003355786-4

1.2). Hence, according to the Third law, the temperature of the magnetic subsystem of diamagnetics is zero as well. This is consistent with the facts of temperature independence of magnetic susceptibility and the absence of the magnetocaloric effect in diamagnets.

Compliance with the laws of thermodynamics is a necessary condition of validity of any physical result whether it is theoretical or experimental. An important advantage of thermodynamics is that it allows one to predict/explain macroscopic properties even if underlying microscopic properties are not well understood.

4.1 THERMODYNAMIC POTENTIALS

Thermodynamics operates with thermodynamic potentials or functions of state whose infinitesimal increment is a total differential of the state parameters. In thermodynamic equilibrium the potential appropriate for a particular case takes minimum of its possible values. In traditional thermodynamics (mostly theory of heat) the thermodynamic variables for systems consisting of a fixed number of microscopic units (e.g., molecules) are temperature T, pressure P, volume V and entropy S; and the thermodynamic potentials are internal energy $U(S, V)$, Helmholtz free energy $F(T, V)$, Gibbs free energy $G(T, P)$ and enthalpy $E(S, P)$. The different forms of thermodynamic potentials are converted from one to another via Legendre transforms [43]. The choice of an appropriate potential for a particular problem is determined by which two of the four pairs of variables are controllable, and by convenience.

Thermodynamics of magnetic systems differs from traditional thermodynamics due to the fact that the field, one of the state parameters in this case, is a vector quantity. This leads to a dependence of the equilibrium properties of magnetizing bodies on the field direction and body shape. Thermodynamics is primarily applicable to para- and diamagnetics[1]. This includes superconductors provided their properties are free of hysteresis.

Application of the magnetic field induces magnetic moment in the specimen and therefore alters the field both inside and outside it. Respectively, a work done by an external agent (e.g., power supply of the magnet) to magnetize the specimen should be added to the named above thermodynamic potentials. As shown by Guggenheim [2] (see also [20, 16]), the work δW needed to change magnetic induction by $\delta \mathbf{B}$ in the *system*[2] is equal to

$$\delta W = \frac{1}{4\pi} \int \mathbf{H} \delta \mathbf{B} dV \qquad (4.1)$$

where the integral is taken over the system volume, i.e., over the entire space occupied by the field.

---

[1]Applicability of thermodynamics to ferromagnetic materials is restricted to the cases when hysteresis is excluded [2].

[2]The system is the specimen plus the field around it.

TABLE 4.1 Forms of free energy appropriate for specimens of magnetizing materials of different geometries. The bottom line in each section is related to the linear materials. (Reprinted from [16].)

Potential	$\eta = 0$ ($\mathbf{H} = \mathbf{H}_0$)	$\eta = 1$ ($\mathbf{B} = \mathbf{H}_0$)	$0 < \eta < 0$
$F(T, \mathbf{B})$	-	$df = -sdT + \dfrac{\mathbf{H}d\mathbf{B}}{4\pi}$	-
	-	$f = f_0 + \dfrac{B^2}{\mu 8\pi}$	-
$\widehat{F}(T, \mathbf{H})$	$d\widehat{f} = -sdT - \dfrac{\mathbf{B}d\mathbf{H}}{4\pi}$	-	-
	$\widehat{f} = f - \dfrac{BH}{4\pi}$	-	-
	$\widehat{f} = f_0 - \dfrac{\mu H^2}{8\pi}$	-	-
$\widetilde{F}(T, \mathbf{H}_0)$	$d\widetilde{f} = -sdT - \mathbf{I}d\mathbf{H}_0$	$d\widetilde{f} = -sdT - \mathbf{I}d\mathbf{H}_0$	$d\widetilde{F} = -SdT - \mathbf{M}d\mathbf{H}_0$
	$\widetilde{f} = \widehat{f} + \dfrac{H_0^2}{8\pi}$	$\widetilde{f} = f - \dfrac{H_0^2}{8\pi}$	-
	$\widetilde{f} = f_0 + (1 - \mu)\dfrac{H_0^2}{8\pi}$	$\widetilde{f} = f_0 + \dfrac{(1 - \mu)}{\mu}\dfrac{H_0^2}{8\pi}$	$\widetilde{F} = F_0 + \dfrac{(1 - \mu)}{1 - \eta(1 - \mu)}\dfrac{V H_0^2}{8\pi}$

The controllable magnetic field is always the applied field $\mathbf{H}_0$, but depending on the specimen geometry and the field orientation, it can also be either induction $\mathbf{B}$ or intensity $\mathbf{H}$ in the specimen. Respectively, the magnetic energy of the system can be presented in three different forms. This implies that instead of 4 thermodynamic potentials in problems without a field, there are 12 potentials to choose from in the field presence. At first glance, it may seem that the use of thermodynamics for magnetic systems is a cumbersome and complex approach. Fortunately, in reality it is right the opposite.

Thermodynamics in magnetic field is considered in detail in [16]. In Table 4.1 we reproduce the summary table from the second chapter of that book. In the table, the potentials F, $\widehat{F}$ ("F-hat")[3] and $\widetilde{F}$ ("F-tilde") are different forms of the Helmholtz free energy written in terms of B, H and H_0, respectively, in an approximation of incompressible specimen. In this approximation, the mechanical part of the specimen internal energy is neglected, and therefore the Helmholtz and Gibbs free energies are equivalent. For this reason, we will

---

[3]In textbooks on superconductivity $\widehat{F}$ is often referred to as the Gibbs free energy.

simply call them free energy. Lowercase letters f, $\widehat{f}$ and $\widetilde{f}$ designate densities of the corresponding forms of free energy[4].

The free energy $F(T, B)$ is appropriate for the transverse geometry (infinite plate in the transverse field, or $\eta = 1$), $\widehat{F}(T, H)$—for cylindrical geometry (long cylinders or infinite plates in a parallel field, or $\eta = 0$), while $\widetilde{F}(T, H_0)$—for all ellipsoidal bodies in the field of an arbitrary orientation ($0 \leq \eta \leq 1$). After Landau and Lifshitz [20], $\widetilde{F}$ is referred to as the total free energy of the *specimen*. The magnetic part of the total free energy in this case (i.e., in the approximation of incompressible material) is magnetic energy E_m given in Eq. (2.33). E_m includes energy of the outside field created by the magnetized specimen. Due to that, as a rule, the concept of density of the total free energy does not carry physical meaning for specimens with η other than 0 and 1.

4.2   CONDENSATION ENERGY

The condensation energy E_c represents an advantage in the free energy of electrons in the S state (i.e., electrons paired in Cooper pairs) with respect to that in the N state (the same, but unpaired or "normal" electrons) at zero field and constant temperature. The origin of this advantage is the compensation of electron spins in Cooper pairs, which makes the pairs effectively spinless and, therefore, reduces the free energy of such a system[5]. E_c is the principal thermodynamic quantity qualitatively distinguishing superconducting and non-superconducting diamagnetics. It is defined as

$$E_c(T, V) = F_{n0}(T, V) - F_{s0}(T, V) = e_c(T)V, \qquad (4.2)$$

where subscripts $n0$ and $s0$ denote the N and S states at zero field, respectively; and e_c is the condensation energy per unit volume.

The fact that E_c is defined through the free energies at zero field implies that it is an additive quantity independent of the shape of superconductor and, therefore, e_c is a function of temperature only. For the same reason Eq. (4.2) is valid for any form of free energy, be it F, $\widehat{F}$ or $\widetilde{F}$.

In turn, e_c defines the thermodynamic critical field $H_c(T)$ as

$$e_c = \frac{H_c^2}{8\pi} \qquad (4.3)$$

In all well studied superconductors $H_c(T)$ is close to a parabolic function given in Eq. (3.8)[6].

---

[4] For superconductors the free energy density in Table 1 corresponds to f_s in the two-fluid model.

[5] The free energy of an ensemble of spinless particles is always less than the free energy of an ensemble of the same particles with spins [32].

[6] Measurements of H_c require pining free sufficiently massive specimens with well defined demagnetizing factor. For this reason, only relatively small number of superconductors can be called well studied.

Using Table 1 and Eqs. (4.2) and (4.3), one can write a general expression for the total free energy of a superconducting specimen[7] at constant temperature. It has the form

$$\widetilde{F} = \widetilde{F}_{s0} - \int_0^{H_0} \mathbf{M}d\mathbf{H}_0 = \widetilde{F}_n - \frac{H_c^2}{8\pi}V - \int_0^{H_0} \mathbf{M}d\mathbf{H}_0 \qquad (4.4)$$

where the subscript 0 in $\widetilde{F}_n$ is omitted because μ_m of superconductors in the N state is unity[8]. This equation is valid for all superconductors in equilibrium states, including those with inhomogeneous magnetization.

The magnetic moment $\mathbf{M}$ of superconductors in equilibrium states is negative. Therefore, the magnetic energy E_m (defined in Eq. (2.33)) is positive and increases with increasing H_0. Respectively, $\widetilde{F}$, being lesser than $\widetilde{F}_n$, also increases until at a definite critical field H_{cr}[9] it becomes equal to $\widetilde{F}_n$. At $H_0 > H_{cr}$ the S state is no longer favorable and, therefore, at H_{cr} the specimen transitions to the N state. Consequently,

$$-\int_0^{H_{cr}} \mathbf{M}d\mathbf{H}_0 = \frac{H_c^2}{8\pi}V = E_c \qquad (4.5)$$

Hence, the condensation energy E_c can also be considered as the magnetic energy it takes to destroy superconductivity at constant temperature, and H_c—as the field characterizing this energy [136]. Accordingly, e_c is the energy required to destroy superconductivity, i.e., to break electron pairs, per unit volume. In turn, the energy of breaking a single Cooper pair or the bonding energy of a pair e_{cp} is

$$e_{cp} = \frac{H_c^2}{8\pi n_{cp}}, \qquad (4.6)$$

where n_{cp} is a number density of the pairs.

From Eq. (4.5) it follows that if $\mathbf{H}_0$ is parallel to a specimen axis, the area under the graph M vs H_0 equals the condensation energy. At an arbitrary orientation of $\mathbf{H}_0$, it should be broken for components parallel to the axes, and the condensation energy is equal to the sum of corresponding areas for each component.

Expressing $\mathbf{M}$ and $\mathbf{H}_0$ in dimensionless units $\mathbf{M}' = 4\pi\mathbf{M}/VH_c$ and $\mathbf{H}_0' = H_0/H_c$, Eq. (4.5) takes the form

$$-\int_0^{H_{cr}'} \mathbf{M}'d\mathbf{H}_0' = -\int_0^{H_{cr}} \left(\frac{4\pi\mathbf{M}}{VH_c}\right)\frac{d\mathbf{H}_0}{H_c} = \frac{1}{2} \qquad (4.7)$$

---

[7]More precisely, the total free energy of superconducting, i.e., paired, electrons in the specimens.

[8]A feeble para- and diamagnetism in the N state is neglected since they are 4-to-5 orders of magnitude less than diamagnetism of the S state.

[9]H_{cr} is not necessarily the same as H_c.

So, the area under magnetization curve in coordinates M' vs H_0' is equal to $1/2$ regardless on the specimen material, temperature and the field orientation provided that it is parallel to one of the specimen axes. This law is referred to as a rule of $1/2$. One cannot help but notice its analogy with the law of corresponding states in simple liquids [43].

From the point of view of thermodynamics, E_c can be considered as an internal resource acquired by the body at the transition to the S-state, which is consumed with the greatest possible effectiveness [16].

4.3 THERMODYNAMIC PROPERTIES OF TYPE-I SUPERCON-DUCTORS

The necessary condition of an equilibrium state of a superconductor in a magnetic field is the homogeneity of the field intensity $\mathbf{H}$ inside it[10]. On this reason the equilibrium superconducting states are observed only in ellipsoidal singly connected bodies (we call them specimens), in which $\mathbf{H}$ is homogeneous according to the Poisson theorem and the continuity of H_t.

However, the homogeneity of $\mathbf{H}$ does not necessarily mean the homogeneity of $\mathbf{B}$: depending on thermodynamic profitability, the induction can be either homogeneous or inhomogeneous. In the first case the specimens[11] are in the MS in which $\mathbf{B} = 0$ over the entire volume, like in Fig. 3.10. In all superconductors this state is thermodynamically profitable (and therefore takes place) when $H \leqslant H_{c1}$ or $H_0 \leqslant H_{c1}/(1-\eta)$. In the second case the specimens are split for the S and N domains where $\mathbf{B}$ equals 0 and $\mathbf{H}$, respectively. Herewith, the domain walls are parallel to $\mathbf{H}$ due to the continuity of B_n, while $\mathbf{H}$ remains uniform due to the continuity of H_t.

The domain structure (also called as flux structure) in the inhomogeneous states is controlled by the S/N surface tension[12] α (energy per unit area of the S/N interface), defined as

$$\alpha = \frac{H_c^2}{8\pi}\delta, \tag{4.8}$$

---

[10]Inhomogeneous $\mathbf{H}$ would lead to inhomogeneous magnetization $\mathbf{I}$ and, as a consequence, to incomplete cancelation of the field-induced microscopic currents in the specimen bulk (see Sec. 1.2.1). The latter would result in a violation of the time reversal symmetry, which is mandatory for the equilibrium systems.

[11]By default, we imply specimens with a uniform density of Cooper pairs at zero field.

[12]More precisely, in type I superconductors, the domain structure is controlled by competing contributions in the free energy, caused by the S/N surface tension inside the specimen, on the one hand, and by the field distribution and shape of domains near the specimen surface, on the other. In type-II materials, it is controlled by the S/N surface tension and the flux quantization.

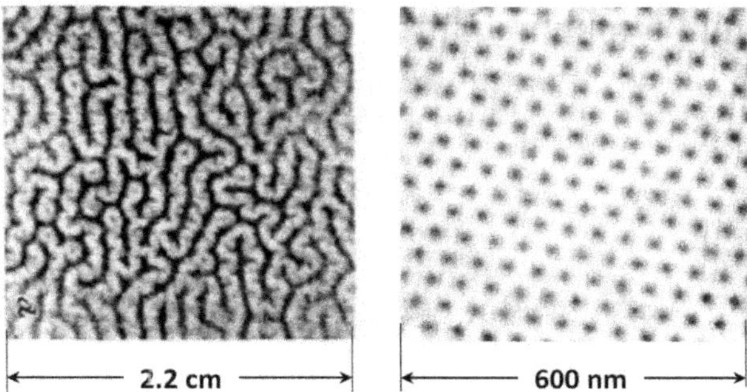

Figure 4.1 Equilibrium domain patterns in the inhomogeneous equilibrium states of type-I (left, the intermediate state) and type-II (the mixed state) superconductors. The left image was obtained on an aluminum plate using powder technique; dark areas are superconducting (After Faber [140]). The type-II superconductor is a single crystal NbSe$_2$ plate; dark circles are normal (After Hess et al. [141]). In both cases the field is directed perpendicular to the plane surface. Reprinted with permission from The Royal Society of London and the American Physical Society.

where δ is so-called domain-wall energy parameter[13] [85]; the surface tension in this form[14] was introduced by Landau [137].

As was mentioned in the previous chapter, the surface tension (and, accordingly, δ) can be either positive or negative[15]. Respectively, superconducting materials are referred to as type-I ($\alpha > 0$) and type-II ($\alpha < 0$) superconductors[16].

The equilibrium inhomogeneous state of type-I superconductors is called the intermediate state (IS); in type-II materials it is referred to as the mixed and Shubnikov state (MXS). Due to inhomogeneous magnetization, the concepts of demagnetizing field and demagnetizing factor are inapplicable to these states. Properties of the IS and MXS are drastically different. Examples of the flux structure in these states are shown in Fig. 4.1. In this section we will discuss the equilibrium properties of type-I superconductors.

---

[13]The rationale for this formula is based on an assumption that the induction at the N/S interface is equal to H_c [137].

[14]The notion of the S/N surface tension was first used by H. London [138].

[15]A possibility of the negative surface tension was mentioned by Ginzburg and Landau [36] and explicitly discussed by Pippard [139].

[16]Such a division was coined by Abrikosov [127].

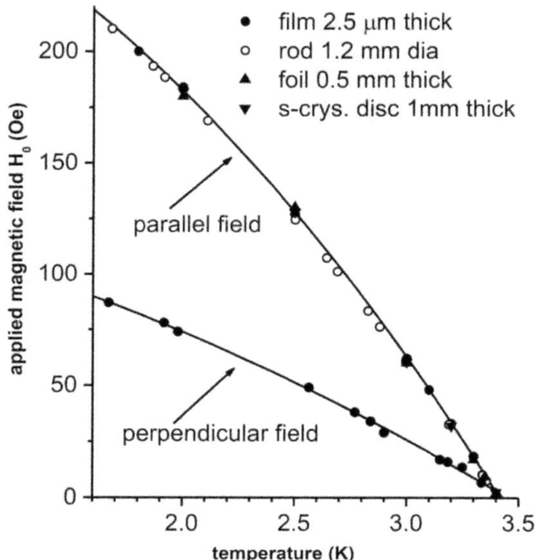

Figure 4.2 Phase diagram of superconducting indium. The data shown by open circles were obtained by Finnemore and Mapother [143]; other data were reported in [144]. All data fitted by the upper curve (parabolic) were obtained on the specimens with $\eta = 0$ (cylindrical geometry). The data fitted by the lower curve were obtained with a specimen of the transverse geometry (the film in a perpendicular applied field). Reprinted from [144] with permission from the American Physical Society.

4.3.1 Phase diagram

Type-I superconductors represent a relatively small group of mostly elementary metals[17].

A typical phase diagram of type-I superconductors is shown in Fig. 4.2. The upper line is the S/N coexisting curve measured using different specimens of cylindrical geometry ($\eta = 0$). Below this curve these specimens are in the MS and above it—in the N state; as we will see in the next subsection, this curve furnishes the thermodynamic critical field $H_c(T)$ determined by the superconducting material. The lower curve fits experimental data on the critical field H_{cr} for a specimen of the transverse geometry (the same film as one of the specimens in the upper plot, but in a perpendicular field). Position of the lower curve depends on the film thickness. For sufficiently thick specimens

---

[17]There is at least one nearly half-by-half alloy BeAu clearly exhibiting type-I superconductivity [142].

(thicker than ≈ 1 mm) it merges with the upper curve. Below the lower curve the specimen is in the IS and above - it is normal. Note that between the curves the specimen of the transverse geometry is in the N state although H_0 is significantly less than H_c[18].

4.3.2 Cylindrical geometry

The specimen-field configuration for the cylindrical geometry is shown in Fig. 1.4a and an example of magnetization curves—in Fig. 3.6. The field intensity $\mathbf{H}$ inside the specimen is equal to $\mathbf{H}_0$ and the phase diagram is represented by the upper curve in Fig. 4.2. At $\mathbf{H}_0 \leq \mathbf{H}_{cr}$ the magnetic permeability $\mu_m = 0$ and, accordingly (see Eq. (1.31)), the specimen magnetic moment $\mathbf{M} = -\mathbf{H}_0 V/4\pi$. Then, using Eq. (4.5), one finds

$$H_{cr} = H_c. \tag{4.9}$$

This equality strictly holds only for type-I superconductors of cylindrical geometry. In such specimens at $H_{cr} = H_c$ superconductivity collapses at once all over the volume meaning that the S/N transition is discontinuous, as it takes place in phase transitions of the first order.

The total free energy (Eq. (4.4)) for this case is

$$\widetilde{F} = \widetilde{F}_n - \frac{H_c^2}{8\pi}V + \frac{H_0^2}{8\pi}V \tag{4.10}$$

and the specimen entropy S_s is

$$S_s = -\left(\frac{\partial \widetilde{F}}{\partial T}\right)_{H_0} = S_n + \frac{H_c V}{4\pi}\frac{dH_c}{dT}, \tag{4.11}$$

where the entropy in the N state S_n does not depend on the field.

We see that the entropy S_s is less than S_n (since $dH_c/dT < 0$) and it does not depend on the field either, because the specimen magnetic moment $\mathbf{M}$ does not depend on temperature[19], in full consistency with experiment (see Fig. 3.6). Therefore, at a fixed temperature $S_s(T, H_c) = S_s(T, 0)$, from which it follows that the entropy of the S state, i.e., the entropy of the ensemble of Cooper pairs, is zero over the entire range of T and H of the superconducting state.

Indeed, S_s is determined by the disorder in microscopic currents and corresponding magnetic moments induced by magnetic field. There is no such currents at $H = 0$, therefore there is no disorder as well, meaning that $S_s(T, 0) = 0$ at all $T < T_c$. Then, from the field independence of S_s the above

---

[18]At crossing the coexisting curves from below, there are traces of superconductivity, which vanish at approximately doubled value of the corresponding critical field (see Sec. 4.5.)

[19]Apply the Maxwell relation $(\partial S/\partial H_0)_T = (\partial M/\partial T)_{H_0}$.

statement follows. In turn, it means that the induced currents are completely ordered.

The zero entropy of the paired electrons is consistent with the Third law, with the absence of thermoelectric effects in superconductors, and with zero magnetic part of entropy in regular diamagnetics. It serves as an additional confirmation of the validity of the zero entropy postulate in the two-fluid model.

Next, from Eq. (4.10) one obtains Keesom's formula for the latent heat of the S/N transition:

$$Q = T(S_n - S_s) = -TV\frac{H_c}{4\pi}\frac{dH_c}{dT} = -\frac{VT}{8\pi}\frac{dH_c^2}{dT}. \tag{4.12}$$

The heat Q is absorbed at the isothermal transition from the S to the N state and released at the transition in opposite direction. The latent heat is a primary characteristic of a phase transition of the first order. Therefore, the S/N transition in type-I superconductors of cylindrical geometry at nonzero field and temperature is indeed the first order phase transition, as was expected based on the data shown in Fig. 3.6.

From Eq. (4.11) it follows that the difference in heat capacities ΔC along the S/N coexisting curve is

$$\Delta C = C_s - C_n = T\frac{dS_s}{dT} - T\frac{dS_n}{dT} =$$

$$\frac{TV}{4\pi}\left[H_c\frac{d^2H_c}{dT^2} + \left(\frac{dH_c}{dT}\right)^2\right]. \tag{4.13}$$

In the upper end of the S/N curve (at $T = T_c$ and $H_0 = 0$) the difference in entropy $\Delta S = S_n - S_s$, the latent heat Q, and the specimen magnetic moment $\mathbf{M}$ vanish. In turn, the heat capacity experiences a jump ΔC_c which, according to Eq. (4.12), is

$$\Delta C_c \equiv \Delta C(T \to T_c) = \frac{T_c V}{4\pi}\left(\frac{dH_c}{dT}\right)^2. \tag{4.14}$$

This is the Rutgers formula [145], originally derived on the basis thermodynamic consideration of the continuous phase transitions proposed by Ehrenfest [146]. This simple formula well agrees with data of Keesom's group shown in Fig. 3.3 [90]. It was confirmed in many other experiments referred in [15].

Thus, the S/N transition at T_c is a continuous phase transition of the second order, similar to those occurring in regular matter, e.g., at the critical point of the liquid-gas phase transition.

However, the picture is different near the lower end of the S/N coexisting curve: at $T = 0$ and $H_0 = H_{c0}$. Here, due to the Third law, $\Delta S = S_n - S_s = 0$ and therefore the slope of the S/N curve is also zero (see Eq. (4.11)); Q and ΔC are equal to zero as well, while $\mathbf{M}$ reaches its maximum. The S/N transition at $T \to 0$ is a quantum phase transition of the first order.

4.3.3 Intermediate state and surface-related magnetic inhomogeneities

There is a widely accepted paradigm stating that in a field lower than H_c type-I superconductors are unstable against formation of the S phase [138]. This is indeed so, but only for specimens of cylindrical geometry. As one can see from the phase diagram in Fig. 4.2, for other specimen/field configurations the situation can be notably different.

Consider a specimen of a non-cylindrical geometry. For example, a long circular cylinder in the field $\mathbf{H_0}$ perpendicular to its longitudinal axis. At low field the specimen is in the MS, i.e., $B = 0$ over the entire volume and therefore $\eta(= 1/2$ for this cylinder) is well defined. Consequently, the field intensity H inside the specimen is equal to $H_0/(1 - \eta)$ (see Eq. (1.29)). Then, due to the continuity of B_n, the outer field bends around the specimen similar as it is shown in Fig. 3.10; and from the continuity of H_t it follows that the external field near the specimen is equal to

$$H_{ext} = \frac{H_0}{(1 - \eta)} \sin \theta, \tag{4.15}$$

where θ is the angle between the normal to the surface and the direction of $\mathbf{H_0}$ (see problem 3.6).

Hence, inside our cylinder $H = 2H_0$ and the external field near an "equatorial line" (the line along the cylinder where $\theta = \pi/2$) also equals $2H_0$. Accordingly, H inside equals H_c when $H_0 = H_c(1 - \eta)$, which is $H_c/2$ for this cylinder. What happens with a further increase of H_0?

On the one hand, the specimen remains superconducting (since $E_m < E_c$) but, on the other hand, it cannot continue being in the MS because in such case the critical field H_{cr} would be equal to $H_0 = H_c\sqrt{1 - \eta}$. If so, the specimen would become "normal" at $H_0 < H_c$, which is impossible for the uniformly magnetized specimen[20]. A scenario in which superconductivity is gradually suppressed on the sides, as in Fig. 4.3a, is also impossible, since such suppression would lead to a decrease in η and, consequently, to the restoration of superconductivity. As was first recognized by Gorter and Casimir [147], the specimen should split into alternating domains of the S and N phases.

The splitting scenario was confirmed by Shubnikov and Nakhutin [148] via measurements of electrical resistance of a tin ball in various directions relative to the applied field[21], and in many subsequent experiments [150]. In our cylinder the equilibrium domains represent discs perpendicular to the cylinder axis, as shown in Fig. 4.3b [110, 111]. Consequently, the resistance of

---

[20]Although this is possible for a specimen with inhomogeneous magnetization, like the one whose coexistence curve is shown by the lower curve in Fig. 4.2.

[21]It was the last paper on superconductivity of Lev Shubnikov submitted in February 1937. In August of that year 36-year-old Shubnikov, a remarkable personality and clearly outstanding physicist, discoverer of the Shubnikov-de Haas effect, co-discoverer of the Meissner effect, discoverer of type-II superconductivity and the author of many other fundamental works [149], was arrested and in November executed by the soviet secret police.

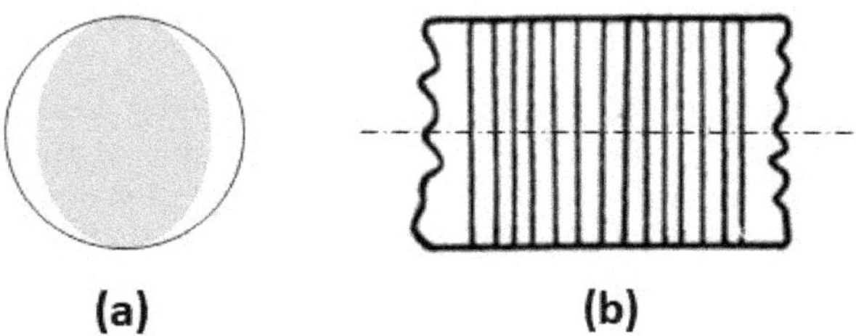

Figure 4.3 (a) A hypothetical (proved impossible) cross-sectional view of a cylinder with partially suppressed superconductivity; the gray area is superconducting. (b) An observed domain structure in a tin cylinder with a weak longitudinal current [110].

the superconducting cylinder is no longer zero in the longitudinal direction, while it is zero in the transverse direction.

As already mentioned, an equilibrium state of type-I superconductors with inhomogeneous induction (or magnetization) is referred to as the intermediate state (IS).

Peierls-London model. A solution of the problem of macroscopic properties of the IS was suggested independently by Peierls [151] and F. London [152, 35]. In the Peierls-London (PL) model the inhomogeneous specimen in the IS is replaced by a homogeneous one with induction averaged over the specimen volume $\overline{B}$. This makes possible to use the demagnetizing factor η and therefore Eq. (1.29) both in the MS and IS. The field intensity H in the IS is assumed constant and equal to its maximum value H_c reached at the critical field H_I of the MS/IS transition. This assumption is based on the fact that the field in N domain is parallel to the domain wall (due to the continuity of B_n) and the critical field for such a geometry equals H_c. On the other hand, this assumption automatically entails that the transition to the N state occurs at $H_{cr} = H_c$.

Magnetic properties of a specimen in the PL model are discussed in [15, 16]. The induction B and intensity H of the field inside the specimen, and the specimen magnetic moment M in this model are

$$0 \leqslant H_0 \leqslant H_c(1-\eta) \begin{cases} B = 0 \\ H = H_0/(1-\eta) \\ M = -H_0 V/4\pi(1-\eta) \end{cases} \tag{4.16}$$

$$H_c(1-\eta) \leqslant H_0 \leqslant H_c \begin{cases} \overline{B} = [H_0 - H_c(1-\eta)]/\eta \\ H = H_c \\ M = -V(H_c - H_0)/4\pi\eta \end{cases} \tag{4.17}$$

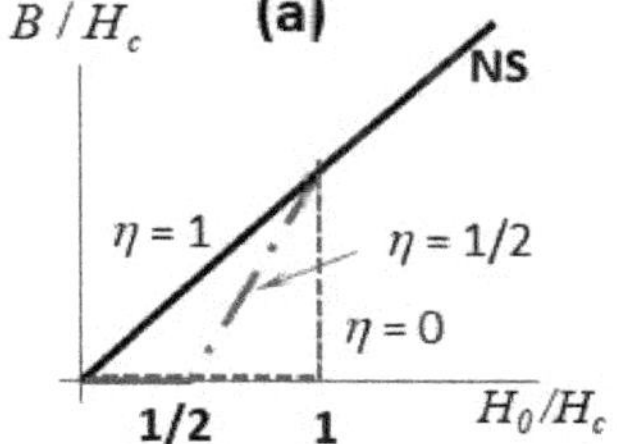

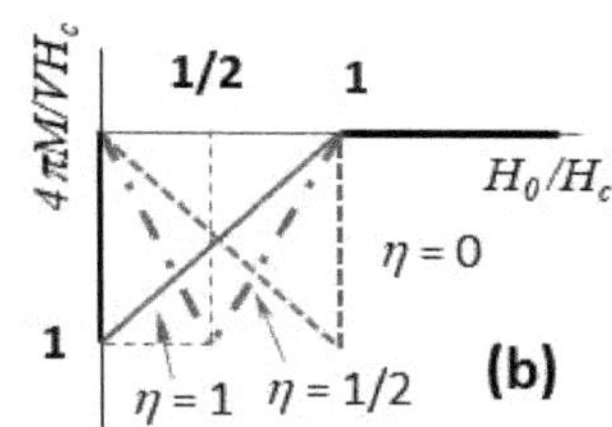

Figure 4.4 The average induction (a) and the magnetic moment (b), both in reduced units, in the Peierls-London model. Solid lines represent these quantities for specimens with $\eta = 1$ (infinite slab in perpendicular field), dash-dotted lines are for $\eta = 1/2$ (long circular cylinder in perpendicular field) and the dashed lines are for $\eta = 0$ (long cylinder in parallel field). NS designates the normal state (solid line at $H_0/H_c > 1$).

Graphs for $\overline{B}$ and M are shown in reduced coordinates in Fig. 4.4. The area under the graphs in Fig. 4.4b is $1/2$. Therefore, the PL model meets the mandatory requirement of thermodynamics. Another important feature is that it is applicable to specimens of all geometries (to singly connected type-I superconductors of any ellipsoidal shape in a field of arbitrary direction). Hence, the PL model represents a global description of the equilibrium properties of such superconductors in an approximation that any possible interactions between the units of real flux structure (i.e., domains) are negligible [16] (see problem 4.2).

Comparison of the PL model with real data shows both a general consistency and discordances. In particular, as one can see from Fig. 4.5a, in cylinders in the transverse field H_I is somewhat bigger than $H_c/2$, and H_{cr} is somewhat less than H_c. Herewith, the differences $\Delta H_I = H_I - H_c/2$ and $\Delta H_{cr} = H_c - H_{cr}$ depend on the diameter of the cylinder: the smaller diameter, the bigger the difference [153, 15]. Similar situation takes place in planar specimens (films) in the transverse field. In this geometry $H_I(= H_0(1-\eta)) = 0$, but, as one can see from Fig. 4.5b, the magnetic moment at $H_0 \to 0$ is greater than $VH_c/4\pi$, and H_{cr} is lesser than H_c. And again, the thinner the film, the greater these differences [30]. However, in both cases the area under magnetization curves meet the rule of $1/2$, as it must be the case for the equilibrium states.

As we will see below, the reason of these discordances is associated with surface-related magnetic inhomogeneous inside and outside the specimen, which are not taken into account in the PL model.

Slab in a transverse field. The first theoretical model addressing the flux structure of the IS was proposed by Landau for a slab in perpendicular field [154, 20]. Assuming that within the slab (a) $H = H_c$ and (b) the domains represent 1D ordered lattice of flat S and N laminae parallel to **H** (and, consequently, parallel to $\mathbf{H}_0$), and (c) the Laplace equation holds everywhere

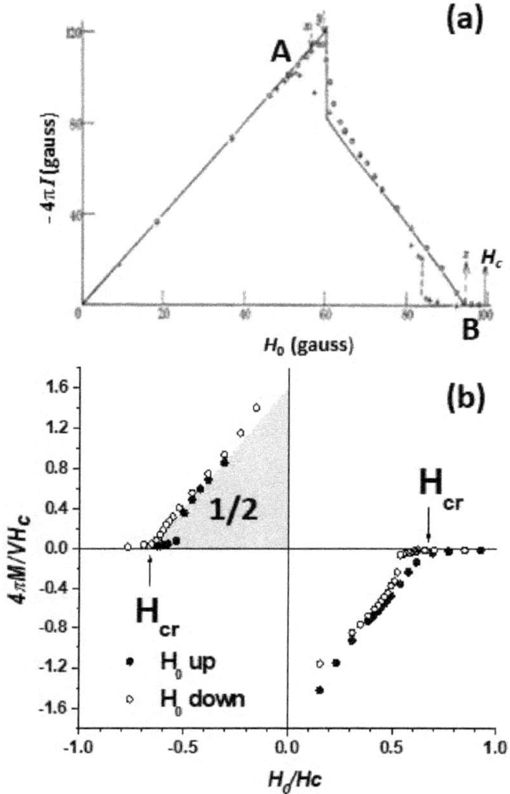

Figure 4.5 Data on equilibrium magnetization in type-I superconductors of different geometry. (a) Cylindrical tin wire of a diameter 0.15 mm in the transverse field at 3.0 K. Dots are the points measured at the increasing field; crosses are the points obtained at decreasing field; solid line is a theoretical curve (after Desirant and Shoenberg [153]). (b) Indium film 3.86 μm thick in the transverse field at 2.5 K (after Kozhevnikov and Van Haesendonck [30]). H_c is the thermodynamic critical field; H_{cr} is the critical field of the S/N transition (in (a) marked as z). In both graphs the area under magnetization curves (0AB in (a)) equals $H_c^2/8\pi$ or 1/2 in reduced units (shown in gray in (b)). A supercooling effect at the N/S transition is well seen in both graphs, and an effect of superheating is well pronounced at the transition from the Meissner to intermediate state in (a). Reprinted with permission from The Royal Society and the American Physical Society.

outside the slab, Landau calculated the domain shape and the field distribution near the surface. Then, taking into account the positive surface tension (Eq. (4.7)) he calculated the widths of the laminae and the period of the laminar structure.

The near-surface field/domain configuration in this model (referred to as Landau's laminar model) is shown in Fig. 4.6a. To satisfy Laplace's equation,

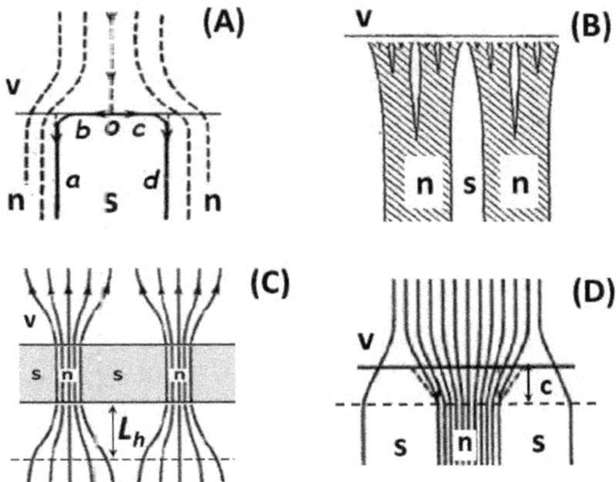

Figure 4.6 Out-of-plane cross-sectional views of the theoretically predicted field distribution and domain shape near the surface of a planar type-I superconductor in a perpendicular field: (A) Landau [154]; (B) Landau [137]; (C) Tinkham [85]; (D) Abrikosov [158]. Letters s and n indicate superconducting and normal phases, respectively; v stands for vacuum. In (C) L_h, the Tinkham healing length, is an effective width of a spacial layer with disturbed field; in (D) c is the size of laminae corners. Note that the Laplace equation is met in all, but Tinkham's scenarios. Reprinted from [159] with permission from Springer Nature.

Landau splits a central field line (in p. o) and bends it for $90°$. Note that this splitting conflicts with the law of flux conservation (see Sec. 1.2.4) while the sharp bend entails huge energy expenditures to maintain such a field distribution.

After critical comments from Peierls, Landau abandoned this model and proposed another one, called branching [137, 155]. In this model, to meet the requirements of Laplace's equation, Landau assumes that laminae branch near the surface as shown in Fig. 4.6b, so the flux enters and exit the specimen being uniformly distributed over the whole surface. To verify this model Landau designed an experiment in which the field distribution is measured in a tiny slot between two hemispheres. Such an experiment was brilliantly performed by Meshkovsky and Shalnikov [156] (see also [15]). The branching model was disproved.

Meshkovsky and Shalnikov were the first to map the flux structure of the IS; for that they scan the specimen surface with a small bismuth's electrical probe, which resistance strongly depends on the magnetic field. Later on, much more detailed images of the flux pattern were obtained with the powder and magneto-optical techniques [150]. It turned out that the equilibrium flux pattern represents a very complicated structure of corrugated laminae, which example is shown in Fig. 4.1. Quite obviously, modeling such a structure is a hopeless task.

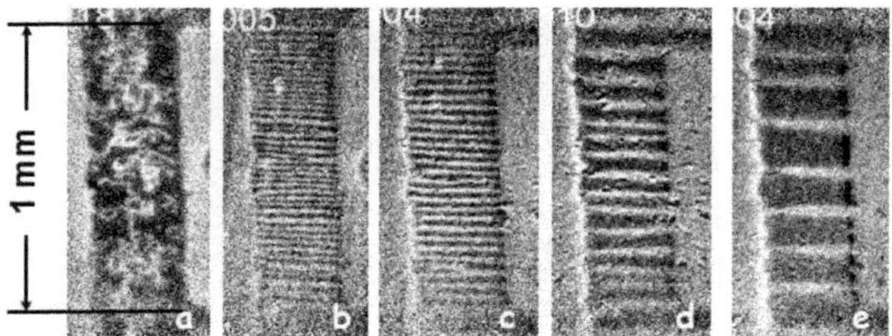

Figure 4.7   Magneto-optical images taken with 2.5-μm-thick indium film at 2.5 K. [$H_{0\parallel}$, $H_{0\perp}$ in Oe]: (a) [0, 1]; (b) [60, 8]; (c) [100, 6]; (d) [110,3]; (e) [115, 1.3]. After Kozhevnikov et al. [144]; reprinted with permission from American Physical Society.

However, as was first demonstrated by Sharvin [157], the laminae can be straitened by tilting the applied field. Another way to straighten the laminae is to apply dc current (see Video 2 in the Support Materials). Examples of regular flux structures in the tilted field are shown in Fig. 4.7.

A laminar IS model for a slab in tilted field (LMTF) was proposed in [144]; its final form is discussed in [16]. The transversely applied field in the model is a limiting case of the tilted one when the in-plane component $\mathbf{H}_{0\parallel} \to 0$. A so-called healing length approximation suggested by Tinkham [85] is adopted for the field distribution near the surface. In this approximation the near surface roundness of laminae is neglected and an effective width of the outer spatial layer with the disturbed field, called the healing length L_h, is taken as

$$\frac{1}{L_h} = \frac{1}{D_n} + \frac{1}{D_s}, \tag{4.18}$$

where D_n and D_s are the widths of N and S laminae, respectively.

The out-of-plane cross-sectional view of the field distribution and shape of domains in this approximation is shown in Fig. 4.6C.

According to the LMTF, all properties of the intermediate state, such as the period of structure $D = D_n + D_s$, the induction in N laminae, the specimen magnetic moment, and the critical field H_{cr} depend on a ratio of the domain-wall parameter δ to the slab thickness d. For instance, the induction in N domains is

$$B^2 = H^2 = H_\perp^2 + H_\parallel^2 = H_c^2 - 4H_c H_{0\perp}\sqrt{\frac{\delta}{d}} \tag{4.19}$$

and the out-of-plane component of the critical field is

$$H_{cr\perp} = \sqrt{H_c^2\left(4\frac{\delta}{d}+1\right) - H_{0\parallel}^2} - 2\sqrt{\frac{\delta}{d}}, \tag{4.20}$$

where $H_{0\perp}$ and $H_{0\parallel}$ are out-of-plane and in-plane components of the applied field, respectively.

As one can see, the thinner the slab, the greater the modeled properties deviate from those in the PL model. And other way around, in the limit $(\delta/d) \to 0$, the LMTF converts into the PL model. For the pure transverse field, the LMTF converts to the model suggested by Tinkham [85]. Hence, measuring properties of a sufficiently thin film in either tilted or transverse filed one can calculate $\bar{\delta}$, an important microscopic parameter related to the size of Cooper pairs (see Sec. 5.2.6).

Comparison of the LMTF with measured data yielded a good fit for some properties and not so good for others [144, 30]. In particular, the decrease of induction with increasing $H_{0\perp}$ (Eq. (4.18)) is consistent with the induction measured by μSR [27], and the decrease of H_{cr} in Eq. (4.19) is consistent with data shown in Fig. 4.5. However, these consistencies are sooner qualitative than quantitative. The discordances were attributed to oversimplified description of the magnetic inhomogeneities near the surface.

Field distribution and domains' shape. An advantage of Tinkham's healing length approximation is its consistency with observed structure in the tilted field. However, it obviously conflicts with Laplace-based magnetostatics, according to which neither voids nor extremums are possible in free space. Recall that the main vector of Landau's efforts was aimed to prevent the appearance of voids in his models.

One more scenario for the field/domain configuration at the transversely applied field was suggested by Abrikosov [158]; it is shown in Fig. 4.6D. According to Abrikosov, all disturbances take place beneath the surface at the depth c calculated on the same formula as the healing length L_h (Eq. (4.18)); in this case, the field near the surface is uniform like in the branching model of Landau. However, this scenario also conflicts with experiments.

The problem of the near-surface field distribution and the shape of domains was addressed experimentally using the low-energy μSR spectroscopy in [159]. Experiments were performed on two high-purity indium films, subjected to a transversely applied field. The depth resolved measurements of the induction, volume fractions of the S and N phases, and the muon depolarization rate were carried beneath the surface (down to 120 nm) and above it (up to 830 nm). For the outside measurements muons were implanted into layers of nitrogen deposited *in situ* on the cold films.

It was found that Tikham's scenario of the domain-filed configuration (Fig. 4.6C) is consistent with the experimental data measured at low field $(H_0 \lesssim H_{cr}/2)$. However, at higher fields the measured configuration differs from the expected scenarios. Specifically, the width of the S domains widens with the decreasing depth instead of expected narrowing. Respectively, the field lines emerge from the N domains converging and the field above them passes through a maximum before it finally relaxes to the uniform applied field far away from the surface.

The measured field/domain configuration is schematically shown in Fig. 4.8. Note the thermodynamic advantage of this configuration: widening of the S laminae lessens the free energy (the gain); at the same time, expanding

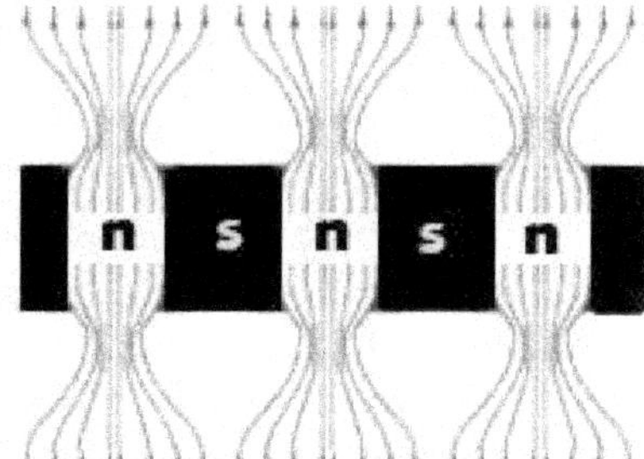

Figure 4.8 The field distribution and domain shape near the surface of a plane specimen in the IS at high field. N and S stand for normal and superconducting phases, respectively. After Kozhevnikov et al. [159]; reprinted with permission from Springer Nature.

the outer layer with the disturbed field makes the transition to the uniform H_0 smoother, hence reduces the losses.

However, both observed field/domain configurations are grossly conflict with the Laplace equation and following from it Earnshaw's theorem, which state that in free space the field lines can converge or diverge only monotonically [20].

This seemingly intractably controversial situation has, however, a straightforward explanation. The space near the specimen in the IS belongs to the specimen and therefore it is *not free*. The field energy in this space, effectively accounted for by the healing length, is a composite contribution into the specimen total free energy. The latter is minimized accounting for all other contributions as well. Consequently, the Laplace equation is not applicable near the specimen; in reported experiments, the width of this region exceeds the specimen thickness.

4.4 THERMODYNAMIC PROPERTIES OF TYPE-II SUPERCONDUCTORS

In type-I superconductors, due to the positive S/N interphase energy, the domain flux structure is thermodynamically unfavorable. Accordingly, in the absence of demagnetizing field H_d (which increases H in diamagnetics) such superconductors stay free of the N domains till the entire condensation energy is exhausted. This case takes place in specimens with $\eta = 0$ in which, as we know, $H_{cr} = H_c$. In other cases (i.e., in the IS), when type-I superconductors have no choice but to allow the flux passing through, they show literally miracles of ingenuity to minimize the losses caused by the surface tension (see Fig. 4.8).

Type-II superconductors behave differently. Having S/N interphase energy negative, they welcome N domains starting from a definite $H = H_{c1}$ called the lower critical field. The latter cannot be zero due to the flux quantization, but it is smaller than H_c (see Secs. 5.2.5 and 5.2.6). This means that the transition to the state with inhomogeneous magnetization takes place in specimens of all geometries, including the cylindrical one. In turn, the mandatory condition Eq. (4.5) dictates that the specimen of cylindrical geometry has to have two critical fields. The second one, referred to as the upper critical field H_{c2}, is greater that H_c.

As mentioned, the equilibrium state of type-II superconductors with inhomogeneous magnetization is referred to as the mixed or Shubnikov state[22]. We abbreviate it as MXS.

To save the condensation energy, the N domains in the MXS are arranged is such a way that their lateral (i.e., S/N) surface area is the largest possible and, therefore, the flux in each of them is the smallest.

The smallest possible flux passing through a superconductor is the superconducting flux quantum $\Phi_0 = ch/2e = c\pi\hbar/e = 2.07 \cdot 10^{-7}$ G cm^2, where h is the Plank constant and $\hbar = h/2\pi$ (see Sec. 5.2.5). Therefore, the N domains in type-II superconductors are identical flux lines carrying Φ_0 each. In literature, these flux lines are often called Abrikosov vortices, discussed in Sec. 5.2.8. Due to the small value of Φ_0 and huge number of the lines passing through the specimen at H_0 just slightly exceeding H_{c1}[23], the lines enter and exit the specimen with a minor field distortion[24].

A qualitative scenario of such kind (without using the flux quantum discovered decade later) was first proposed by Pippard [139] to explain superconducting properties of thin films. As was later shown by Abrikosov [127], it is also applicable to bulk specimens of superconducting alloys, whose magnetic properties were reported by Shubnikov with coworkers twenty years earlier [126]. As mentioned, that study marked the discovery of type-II superconductivity [149].

When the applied field increases at constant temperature the total flux Φ passing through the specimen in the MXS changes with the small steps $\Delta\Phi = \Phi_0$. Therefore, contrarily to type-I materials, the S/N transition at H_{c2} is a continuous thermodynamic phase transition of the second order, as it was first found in Kharkiv (see Fig. 4.9).

Due to the minor distortion of the field near the surface, one can expect that $H_{c2}(= H_{cr})$ does not depend on the shape and dimensions of the specimen. As we will see, experiment confirms this expectation.

---

[22]Sometimes it is also called the vortex state, but we will avoid this term as it can be misleading (see Sec. 5.2.8)

[23]To appreciate this fact, the number density of the flux lines passing through a superconductor of the transverse geometry in the Earth field (≈ 0.5 G) is 10^5 mm^{-2}.

[24]Note that the field distribution near the surface of type-II superconductors is close to that, which Landau sought to construct in his branching model for type I superconductors (see Fig. 4.6B).

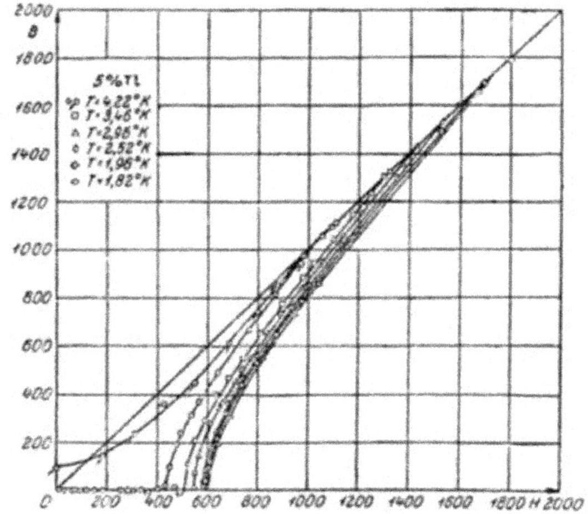

Figure 4.9 Induction B (G) vs field intensity H (Oe) in a Pb-5wt%Tl alloy reported by Shubnikov et al. [126]. The induction was computed from magnetization measured by the ballistic technique (see Sec. 3.1.3) on a ZFC cylindrical specimen (5 mm in diameter and 60 mm in length) in the parallel field.

Next, since the N domains communicate (interact) with each other through the disturbed external field near the surface, the minor field distortion implies that the flux lines in the MXS can be considered not-interacting[25]. This is indeed what is observed experimentally [160].

After all, the absence of interaction between the flux lines and their identity means that in the plane transverse to **H** the lines form 2D hexagonal lattice due to rotational symmetry in this plane. The observation of a hexagonal lattice structure of flux lines was first reported by Essmann and Träuble [161] and confirmed in many subsequent experiments [150]. An image of the flux lattice obtained using a scanning-tunneling microscope [141] is shown in Fig. 4.1.

4.4.1 Magnetization curves and the phase diagram

Type-II superconductors represent a vast majority of superconducting materials, most of which are alloys and multicomponent compounds. Due to that it is a tremendous challenge to fabricate specimens with virtually no pinning, which are necessary for measuring equilibrium properties. There are, however,

---

[25]The interaction of the flux lines in the balk would require setting additional currents around them. But since such currents would (a) increase the specimen free energy and (b) violate the time-reversal symmetry, they are not possible.

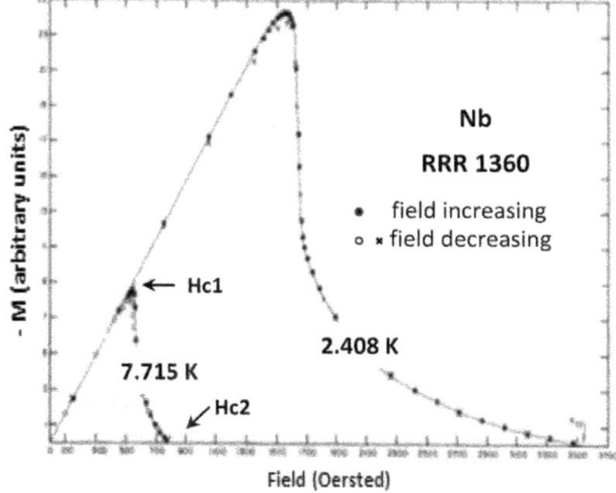

Figure 4.10   Reversible magnetization curves of a cylindrical niobium specimen (0.5 in. long and 0.02 in. in diameter) in the parallel field. RRR (residual resistivity ratio) indicates on an extremely high purity of the specimen bulk. After Finnemore et al. [162]; reprinted with permission from the American Physical Society.

two type-II materials, niobium and vanadium, which, due to their elemental composition are appropriate for fabrication of specimens with minimal pinning. Nowadays the purest of available type-II superconductors is niobium.

Typical reversible magnetization curves for type-II superconductors of cylindrical geometry are shown in Fig. 4.10. At $H \leq H_{c1}$ the specimen is in the Meissner state. in which $I = -H/4\pi$. In the MXS (between H_{c1} and H_{c2}), the magnitude of $I(H)$ decreases non-linearly. Herewith, as first noted by Finnemore et al. [162], the curve shape in the MXS depends on temperature: it approaches linear when T increases. The temperature dependence of thermodynamic critical field H_c (computed using Eq. (4.5) with $H_{cr} = H_{c2}$) follows the same parabolic law as in type-I superconductors (Eq. (3.8)).

Relatively recently, owing to the progress in fabrication of high-purity niobium films [163], the magnetization curves were measured in specimens of the transverse geometry [160]. As we already know, in this geometry the MS is absent and, therefore, the MXS starts immediately from $H_0 \to 0$. Typical data obtained are shown in Fig. 4.11. Over the entire field range up to $H_{c2\perp}$ (H_{c2} in the transverse field) magnetization is the linear function of H_0. Herewith the area under the curve is the same as that measured in the parallel field, as it must be the case for the equilibrium state, and $H_{c2\perp}$ is equal to H_{c2} measured in parallel field.

A typical phase diagram of type-II superconductors of cylindrical geometry is shown in Fig. 4.12. Below the curve $H_{c1}(T)$ the specimen is in the MS. Same as in type-I superconductors, position of this curve depends on the specimen geometry; in the transverse geometry it coincides with the temperature

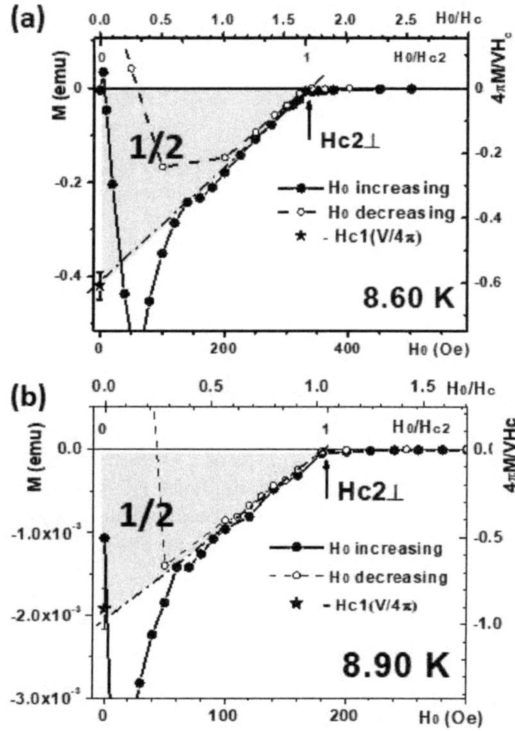

Figure 4.11 Magnetization curves of niobium specimens of the transverse geometry. (a) 1 mm thick single crystal disc 7 mm in diameter; (b) 5.7 μm thick film with RRR = 640. Solid and open circles are data measured at ascending (ZFC) and descending (FC) applied field H_0. The dash-dotted line is $M(H)$ following from the AMMS. In both graphs 1/2 is the area (shown in gray) under magnetization curve in reduced coordinates $4\pi M/VH_c$ vs H/H_c representing the condensation energy $H_c^2 V/8/\pi$ with H_c inferred from the magnetization curves measured in the parallel field. The star is magnetic moment following from the AMMS as indicated. H_{c2} and $H_{c2\perp}$ are the upper critical field measured in parallel and perpendicular fields, respectively [160]. Reprinted with permission from Springer Nature.

axis. The MXS takes place between H_{c1} and H_{c2} curves. Above H_{c2} there are traces of superconductivity in form of filaments, whose density vanishes at H_{c3} (see Sec. 4.5). The locus of $H_c(T)$, $H_{c2}(T)$ and $H_{c3}(T)$ are the same for all geometries of the ellipsoidal bodies of the material under consideration. In specimens of cylindrical geometry H_{c1}, H_c, H_{c2} and H_{c3} are intrinsic parameters of the material. The critical fields are related: H_c is close to the geometrical mean of H_{c1} and H_{c2} [85]; $H_{c3} \lesssim 2H_{c2}$.

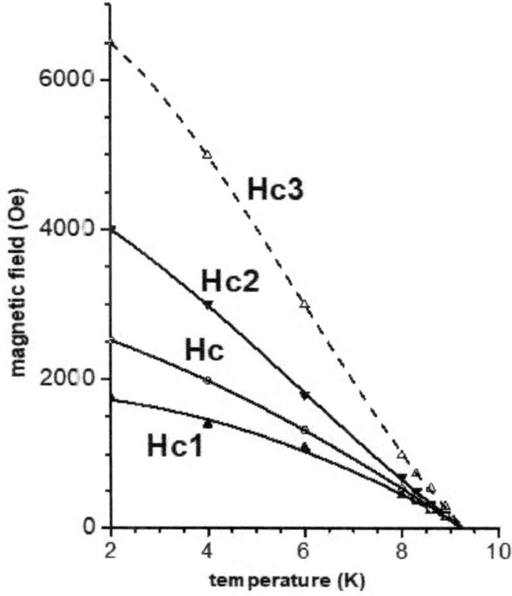

Figure 4.12   Phase diagram of type-II superconductors. Shown diagram is acquired from magnetization curves of a niobium film (5.7 μm thick, RRR = 640) measured in the parallel field. The field on the vertical axis is the applied field H_0 which in this geometry equals to the field intensity H inside the film. The curves $H_{c1}(T)$ and $H_{c2}(T)$ represent the lower and upper critical fields; $H_{c3}(T)$ is the critical field of nucleation of superconductivity; and $H_c(T)$ is the thermodynamic critical field retrieved from the area under the magnetization curves [160]. Reprinted with permission from Springer Nature.

4.4.2   Averaged model of the mixed state (AMMS)

The Meissner state in type-II superconductors does not differ from this state in type-I materials. On the other hand, as shown in the next section, super-conductivity above H_{c2} takes a very small fraction of the condensation energy. Therefore, the main task of a theory of thermodynamic properties of type-II superconductors is the description of the mixed state. In most textbooks, its solution comes down to a fragmentary description of the magnetization curve of cylindrical specimens of extreme type-II superconductors (those for which the GL parameter $\kappa \gg 1$) with the use of the GL or London theory modified to account for the flux passing through Abrikosov vortices. In all cases the fact that within the specimens $H = H_0$ is utilized (see, e.g., [85, 158, 164]).

There was an attempt to compose a model similar to the PL model for the IS [164]. However, it was unsuccessful due to lack of information about the field intensity H in other than cylindrical specimens. The measurement of the magnetization curve for the transverse geometry [160] made it possible to resolve this problem. The model, called the averaged model of the mixed state (AMMS) is discussed in detail in [160] and [16]. Its compressed summary is as follows.

Similar as in the PL model, in the AMMS the specimen with inhomogeneous induction B is replaced by a specimen of the same shape with the induction averaged over the volume $\overline{B}$. Such a replacement makes the demagnetizing factor a well-defined quantity, which allows using Eq. (1.29). The rest directly follows from the experimental results on magnetization shown in Fig. 4.11 (see [16, 160] for details).

Applying H_0 parallel to a specimen axis, with respect to which the demagnetizing factor is η, the intensity H within this specimen is

$$H = H_{c1} + \frac{H_{c2} - H_{c1}}{H_{c2} - H_{c1}(1 - \eta)}[H_0 - H_{c1}(1 - \eta)]. \tag{4.21}$$

The averaged induction, calculated with the use of Eq. (1.29), is

$$\overline{B} = \frac{H_{c2}}{H_{c2} - H_{c1}(1 - \eta)}[H_0 - H_{c1}(1 - \eta)]. \tag{4.22}$$

And the specimen magnetic moment M calculated from the definition of H (Eq. (1.13)) is

$$\frac{4\pi M}{V} = -H_{c1} + \frac{H_{c1}}{H_{c2} - H_{c1}(1 - \eta)}[H_0 - H_{c1}(1 - \eta)]. \tag{4.23}$$

Graphs for these quantities are shown in Fig. 4.13.

As one can see from Fig. 4.13d, at $H_0 = H_{c1}$ the magnetic moment of cylindrical specimens ($\eta = 0$) equals $-H_{c1}V/4\pi$, and at $H_0 = H_{c2}$ the moment is zero in specimens of all geometries. On the other hand, the area above the magnetization curves for different η is the same, implying that the condensation energy does not depend on the field orientation and therefore the model meets the rule of $1/2$.

If one converts material from type-II to type-I (in which by definition $H_{c1} = H_{c2} = H_c$) moving point b toward point c in Fig. 4.13b, the graphs convert to the graphs in Fig. 4.13a, and Eq. (4.20) yields $H = H_c$. Therefore, the PL model represents the limiting case of the AMMS for type-I superconductors.

From the graphs in Fig. 4.13d and Eq. (4.5) it follows that (see Problem 4.4)

$$-\int_0^{H_{c2}} \frac{4\pi \mathbf{M}}{V}d\mathbf{H}_0 = \frac{H_{c1}H_{c2}}{2} = \frac{H_c^2}{2}. \tag{4.24}$$

Hence, H_c is a geometrical mean of H_{c1} and H_{c2}. This is an empirical rule known for extreme type-II superconductors [85]. The AMMS extends this rule for all type-II materials. Note the consistency of Eq. (4.24) with the physical

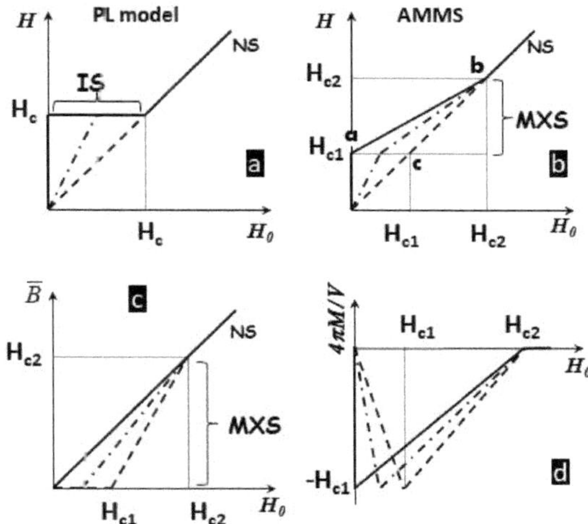

Figure 4.13 The field intensity H, average induction $\overline{B}$ and magnetic moment M vs applied field H_0 in specimens of different geometries. (a) H in the PL model for type-I superconductors; (b) H in the AMMS for type-II superconductors; (c) and (d) $\overline{B}$ and M (in the units of $4\pi M/V$) in the AMMS, respectively. Solid lines depict these quantities for specimens of the transverse geometry ($\eta = 1$); dash-dotted lines are for long circular cylinder in perpendicular field ($\eta = 1/2$), and the dashed lines are for specimens of cylindrical geometry ($\eta = 0$). IS, MXS and NS stand for the intermediate, mixed and normal states, respectively [160]. Reprinted with permission from Springer Nature.

meaning of condensation energy as the magnetic energy needed to destroy superconductivity.

The solid line in Fig. 4.13d represents the magnetization curve in AMMS for the transverse geometry. One can see that, the AMMS is in quantitative agreement with the experimental data for this geometry. However, consistency with the experiment worsens with decreasing η. Experimental magnetization curve in the MXS for cylindrical geometry (see Fig. 4.10) is not linear, whereas in the model $M(H_0)$ is the straight line. There are different approaches to explain this non-linearity (see, e.g.,[85, 165, 127]). In most of them the lateral surface of the cylindrical specimens plays a central role (note that the concentration of defects is highest near the surface), but the temperature dependence of the shape of the M-curve is ignored. Nowadays this question is still open, but the fact that in the transverse geometry this surface is absent is likely a factor explaining the agreement of the AMMS with the experiment.

In [160] it is shown that AMMS along with the experimental magnetization curve for the transverse geometry provides the basis for a number of fundamental conclusions about so-called vortex matter. In particular, this furnishes an experimental proof that the flux lines do not interact and, therefore, a frequently used concept of interacting Abrikosov vortices is questionable. Recall that the absence of interaction between the flux lines follows from thermodynamics. We will come back to this point in Sec. 5.2.8.

4.5 NUCLEATION OF SUPERCONDUCTIVITY. THE CRITICAL FIELD H_{C3}

Systematic studies of magnetic properties of type-II superconductors were started in 1960s (see, e.g., [143]). It was revealed that H_{cr} determined from resistivity measurements differs quite significantly (about for a factor of 2 and even greater) from that determined from magnetization data. Saint-James and De Gennes [166] based on the GL theory proposed that this difference is due to the influence of the specimen surface on the nucleation of superconductivity.

According to Saint-James and De Gennes, in specimens of cylindrical geometry (e.g., a plate in parallel field) superconductivity nucleates near the surface at $H_{c3} = 1.7H_{c2}$. However in specimens of the transverse geometry (the same plate in perpendicular field) the nucleation occurs in the bulk at H_{c2}. Between H_{c2} and H_{c3} superconducting phase in cylindrical specimens forms a thin continuous surface sheath. The latter is not visible in magnetization measurements due to limited sensitivity of magnetometers available at that time. However, this sheath short-circuits the potential leads at measurements of resistivity. Hence, in cylindrical specimens the S/N critical field determined through magnetization measurements is H_{c2}, whereas the critical field following from the resistance measurements is H_{c3}.

In spite of its general acceptance (see, e.g., [167, 85]), such an interpretation of superconductivity nucleation has left many questions unanswered. For instance, by definition, the field passes through specimens in the N state being unperturbed, i.e., not noticing the surface. Then, why does the field of superconductivity nucleation depend on the field-to-surface orientation?

The problem of superconductivity above H_{c2} was revisited in a series of experiments with high-purity niobium specimens in both parallel and transverse fields using the state of the art techniques for measuring magnetization, electrical resistance and induction plus the scanning Hall probe microscopy [168]. It was shown that superconductivity nucleates in the specimen bulk at $H_{c3} \approx 2H_{c2}$ regardless on the field orientation; between H_{c2} and H_{c3} the superconducting phase forms filaments parallel to **H**. Under increasing field the filament number density decreases vanishing at H_{c3}.

Traces of superconductivity above H_{cr} are also take place in type-I superconductors. In particular, they are well seen at ascending field in μSR probing of the IS [27] and in zooming up the graphs in Fig. 3.6 for the MS. At descending field these traces are absent due to supercooling of the N phase caused by positive S/N surface tension.

Volume of superconducting phase and corresponding amount of the condensation energy associated with superconductivity above H_{c2} was estimated from the area under magnetization curves reported in [168]. This yielded $\sim 1\%$ of the total condensation energy.

4.6 PROBLEMS

4.1. The entropy difference $\Delta S = S_s - S_n$ in a cylindrical type I superconductor is given by Eq. (4.11). At all temperatures $0 < T < T_c$ it is negative. However at $T = 0$ entropy in both states is zero due to the Third law, and at $T = T_c$ the difference ΔS is also zero by definition of T_c. Therefore, $\Delta S(T)$ has a maximum at some temperature T_1.

Find: (a) What is the value of T_1? (b) What happens to the difference in heat capacities $\Delta C = C_s - C_n$ at this temperature? (c) How does ΔC depend on T below T_1? Use Eq. (3.8) for $H_c(T)$.

Solution. (a) From (4.11) the derivative $d\Delta S/dT$ is

$$\frac{d\Delta S}{dT} = \frac{V}{4\pi}\left[\left(\frac{dH_c}{dT}\right)^2 + H_c\frac{d^2H_c}{dT^2}\right]$$

Hence, $T_1 = T_c/\sqrt{3} \approx 0.6T_c$.

(b) The difference in heat capacities

$$\Delta C = T\frac{d\Delta S}{dT} = V\frac{H_{c0}^2}{2\pi T_c^2}\left[3\frac{T^3}{T_c^2} - T\right].$$

Hence, at $\Delta C = 0$ at $T = T_1$ and $T = 0$.

On the other hand,

$$C_s = \Delta C + C_n = \Delta C + \alpha T,$$

where α was calculated in Problem 3.1. Plugging it in here we see that $C_s(T)$ is a cubic parabola.

(c) The derivative $d\Delta C/dT$ is

$$\frac{d\Delta C}{dT} = \frac{d\Delta S}{dT} + \frac{VT}{4\pi}\left[\frac{d^3H_c}{dT^3} + 2\frac{dH_c}{dT}\cdot\frac{d^2H_c}{dT^2}\right]$$

From here it follows that $\Delta C(T)$ has a minimum at $T_2 = T_c/\sqrt{5} \approx 0.45T_c$.

The calculated dependencies for $\Delta S(T)$ and $\Delta C(T)$ are consistent with experimental results (see [15] for references).

4.2. Consider a superconducting slab in perpendicular field as shown in Fig. 4.14. Dimensions of the slab $L_x \sim L_y \gg d$. Assume that flux passing the slab forms a 1D lattice with period D of the N and S domains which widths are D_n and D_s, respectively. Induction B in N and S domains equals to H_c and zero, respectively. Neglect the near-surface distortions of the field and

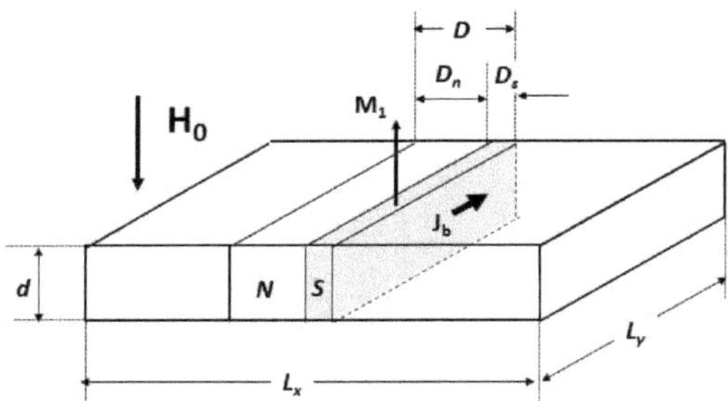

Figure 4.14 Schematics of the slab specimen in problems 4.2 and 4.3. N and S are the normal and superconducting laminae assumed of the shape of rectangular parallelepipeds with widths D_n and D_s, respectively. $D = D_n + D_s$ is a period of the flux structure. J_b is the surface current formed by bound currents induced in Cooper pairs. M_1 is the magnetic moment of a single S domain. The slab thickness d is much lesser than its lateral dimensions L_x and L_y. $\mathbf{H}_0$ is the applied field.

domains' shape. Show that magnetic properties of this slab (M, B and H) are identical to those of a specimen with $\eta = 1$ in the PL model.

Solution. Referring to Fig. 4.14 and using the boundary condition for B_n Eq. (1.25) we write

$$M_1 = \frac{(L_y D_s) J_b}{C} = -\frac{L_y D_s}{c} \frac{c(B-0)}{4\pi} d = -\frac{L_y D_s}{c} \frac{cH_c}{4\pi} d$$

$$M = \frac{L_x}{D} M_1 = -\frac{H_c}{4\pi}(L_x L_y d)\frac{D_s}{D} = -\frac{H_c}{4\pi} V \rho_s$$

$$\rho_s \equiv \frac{V_s}{V} = \frac{D_s}{D} = 1 - \rho_n = 1 - \frac{D_n}{D} = 1 - \frac{H_0}{H_c}$$

the last expression follows from the flux conservation. So,

$$M = -\frac{V H_c}{4\pi}(1 - \rho_n) = -\frac{V}{4\pi}(H_c - H_0).$$

Note that M does not depends on the specific pattern of the flux structure, but only on the total volume of the N phase $V_n = \rho_n V$; therefore, domains can be moved while keeping constant V_n. This implies that the domains can be either ordered, like in Fig. 4.14, or disordered, like in Fig. 4.1, or they can be collected all together in one place. The magnetic moment of all these systems will be the same as soon as energy contribution due to surface related disturbances can be neglected, or if $d \gg \delta$.

Now, due to the boundary condition for H_t (1.20)

$$H = H_c.$$

And from (1.13)

$$\overline{B} = H + \frac{4\pi M}{V} = H_0.$$

These quantities (M, H, and B) are the same as in the PL model (Eqs. (4.17)).

4.3 For the specimen of the previous problem find an expression for the the total free energy $\widetilde{F}(H_0)$ and calculate M.

Solution. Referring to Table 4.1 and Fig. 4.14 and accounting that $\widetilde{F}_{n0} = F_{n0} = F_n$, we write

$$\widetilde{F} = F - \frac{H_0^2}{8\pi} V =$$

$$F_{n0} = F_n - \left(\frac{H_c^2}{8\pi} dL_y D_s\right) N + \left(\frac{H_c^2}{8\pi} dL_y D_n\right) N - \frac{H_0^2}{8\pi} V =$$

$$F_n - \frac{H_c^2}{8\pi} V \rho_s + \frac{H_c^2}{8\pi} V \rho_n - \frac{H_0^2}{8\pi} V.$$

Taking into account that $\rho_n = H_0/H_c$ and $\rho_s = 1 - \rho_n$ we continue

$$\widetilde{F} = F_n + \frac{V H_c^2}{8\pi}(2\rho_n - 1) - \frac{H_0^2}{8\pi} V =$$

$$F_n + \frac{V H_c^2}{8\pi}\left(2\frac{H_0}{H_c} - 1\right) - \frac{H_0^2}{8\pi} V.$$

Note that $\widetilde{F}$ in this model does not contain a free parameter which could be varied to minimize $\widetilde{F}$, i.e., to optimize consumption of the condensation energy. In other words, any flux pattern yields the same free energy, meaning that the system is in neutral equilibrium.

Now the magnetic moment of this specimen is

$$M = -\frac{\partial \widetilde{F}}{\partial H_0} = -\frac{V}{4\pi}(H_c - H_0),$$

M is the same as in the previous problem and in the PL model for $\eta = 1$, as it should be.

4.4. Derive the formula (4.24) for magnetic energy of a type-II superconducting specimen of an arbitrary geometry $(0 \leq \eta \leq 1)$.

Hint: Use M from Eq. (4.16) for $0 \leq H_0 \leq H_{c1}$ and (4.23) for $H_{c1} \leq H_0 \leq H_{c2}$.

CHAPTER 5

SUPERCONDUCTIVITY. ELECTRODYNAMICS

As we saw in the previous chapter, the laws of thermodynamics allow one to explain nearly all equilibrium properties of superconductors. However, thermodynamics cannot (and not supposed to) explain the nature of these properties, let alone the transport characteristics. For that we have to consider the electrodynamics of superconducting media based on microscopic properties. This is the subject of the remainder of this book.

5.1 DEFINITION OF SUPERCONDUCTIVITY

So far we discussed superconductivity without its specific definition. Before discussing electrodynamics we have to first clarify this point.

Historically, superconductivity was seen as a peculiar phenomenon in which the electrical resistance of some metals is zero, similar as in the hypothetical perfect conductor discussed by Maxwell. However, as it became especially clear after the discovery of the Meissner effect, apart from zero resistivity, superconductors have nothing in common with the perfect conductor. Then it was suggested that the electrical properties of superconductors are a consequence of an ideal diamagnetism of these materials. However, a definition of superconductivity as a phenomenon of ideal diamagnetism was proven to be also incomplete because, in particular, the magnetic moment of a superconductor may have both signs (see Fig. 3.7). After all, coming from the success of the two-fluid model, one could try to define superconductivity as a state with ideal order (zero entropy), but then it is necessary to specify the physical meaning of this order (what and how is ordered) and why this leads to zero resistance and zero induction.

A common and unique feature of all superconductors in all (equilibrium and non-equilibrium) states is the presence of Cooper pairs. Consequently,

DOI: 10.1201/9781003355786-5

superconductivity can be defined as a phase state of conducting media in which a statistically significant proportion of conduction electrons is composed of stable Cooper pairs.

Among a broad spectrum of properties, zero resistivity, zero entropy and zero induction of the S phase[1] can be considered as the principal properties of all superconductors. One of the most extraordinary and hardest to recognize facts is that the magnitude of these real physical quantities is not merely very small, but exact zero. (Indeed, what does it mean to measure a quantity of zero magnitude?) These three "big zeros" are on equal footing, meaning that neither one can exist without the other two. Therefore, these zeros have a common origin, and they (all three) should be automatically reproduced in an adequate theoretical model. A model of such kind is considered below.

5.2   MICRO-WHIRLS MODEL

The micro-whirls model (MWM) owes its appearance to an experimental LE-μSR study of type-I and type-II superconductors in the Meissner state [130]. Some results of this study are available in [116]. The model was first published in [116] and in a shortened form in [169].

5.2.1   Cooper pairs

By definition, a Cooper pair represents a bound state of two conduction electrons with opposite spins and zero total linear momentum; kinetic energy of each of these electrons is close to Fermi energy [11, 170]. In other words the Cooper pair is a stable spinless formation of conduction electrons with equal and opposite kinetic linear momentums about their center of mass.

In formal expressions, these conditions are

$$\mathbf{s}_{cp} = \mathbf{s}_1 + \mathbf{s}_2 = 0 \tag{5.1}$$

and

$$\mathbf{p}_{cp} = \mathbf{p}_1 + \mathbf{p}_2 = 0 \tag{5.2}$$

where $\mathbf{s}$ and $\mathbf{p}$ designate the spin angular momentum and kinetic linear momentum, respectively; subscripts 1 and 2 refer to the first and second electrons in the pair, and the subscript cp denotes Cooper pair. Electrons with the Fermi energy are non-relativistic, so $\mathbf{p} = m\mathbf{v}$, where m is the free electron mass and $\mathbf{v}$ is equal (or very close) to the Fermi velocity.

As noted in Sec. 4.2, the condition Eq. (5.1), originally derived from quantum-mechanical reasoning [11], provides a thermodynamic advantage of

---

[1]The S phase can occupy either the entire volume of a superconducting body, as it takes place in the MS, or a part of it in all other cases.

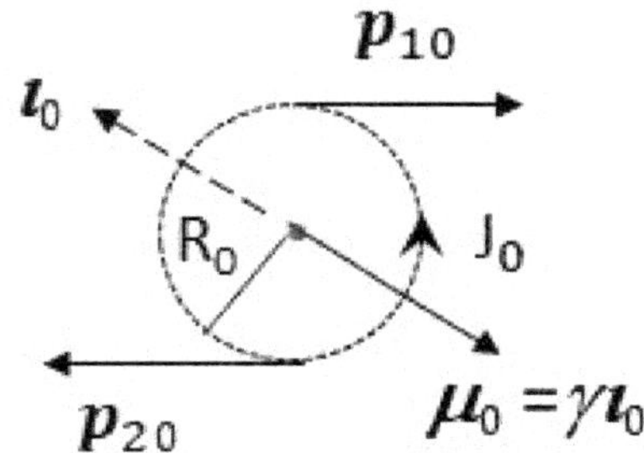

Figure 5.1 Cooper pair at zero field. $\mathbf{p}_{10}$ and $\mathbf{p}_{20}$ are kinetic linear momentums of the paired electrons, the magnitude of which equals mv_F; ι_0 is the pair angular momentum; μ_0 is its magnetic moment; γ is the gyromagnetic ratio; J_0 is the current due to the orbital motion of electrons; and R_0 is the radius of the orbit; a dot in the middle designates the pair center of mass. Spins are not shown since they are mutually compensated.

the S state, since compensation of the spins reduces free energy of the conducting body. The same condition justifies the Cooper pairing as it nullifies the spins.

The condition Eq. (5.2) along with the stability of the pairs[2] means that paired electrons orbit their center of mass. Hence, each pair possesses an orbital angular momentum ι and corresponding magnetic moment $\mu = \gamma\iota$, where γ is gyromagnetic ratio of the orbiting spinless electrons. As we know, g-factor in this case equals unity (see Eq. (2.3)), exactly as it was measured by I. Kikoin and Goobar (see Sec. 2.2).

A sketch of a Cooper pair in a zero field is shown in Fig. 5.1. The orbit diameter $2R_0$ corresponds to the correlation lengths ξ_0 of the BCS theory and ξ of the GL theory, which are close at these conditions. We remind that the pairs overlap[3] but it affects neither stability, nor mobility of the pairs. Taking into account the wave side of the electron nature, this fact is similar as the overlapping of myriads of electromagnetic waves around everyone does not prevent her/him from seeing objects or listening to broadcasts from the other side of the globe.

The pairs stability also means that the electron orbiting occurs without energy dissipation. Therefore, the pairs obey the Bohr-Sommerfeld quantization condition. Herewith, to comply with the law of momentum conservation in a magnetic field, the linear momentums should be taken in a generalized or canonical form. Hence, the quantization condition for a single Cooper pair is

$$\oint \widetilde{\mathbf{p}}_{cp} \cdot d\mathbf{l} = 2\pi R \widetilde{p}_{cp} = 2\pi \widetilde{\iota}_{cp} = nh, \qquad (5.3)$$

---

[2]The stability indicates that $\mathbf{p}_1$ and $\mathbf{p}_2$ cannot lie on one straight line.

[3]As estimated from the uncertainty principle, $\xi_0 \sim 10^{-4}$ cm [11].

where $\widetilde{p}_{cp}$ and $\widetilde{\iota}_{cp}$ are the generalized linear and angular momentums of the pair, respectively; R is the radius of orbital motion of paired electrons; h is the Planck constant; and n is a non-negative integer ($n = 0, 1, 2, ...$).

On a reason, which will be clear below, a theoretical model of superconductivity based on the generalized Bohr-Sommerfeld quantization condition Eq. (5.3) is called the micro-whirls model, abbreviated MWM.

By definition of $\widetilde{\mathbf{p}}$ [31], the generalized linear momentum of the paired electrons $\widetilde{\mathbf{p}}_{cp}$ is

$$\widetilde{\mathbf{p}}_{cp} = \widetilde{\mathbf{p}}_1 + \widetilde{\mathbf{p}}_2 = (m\mathbf{v}_1 + \frac{e\mathbf{A}_1}{c}) + (m\mathbf{v}_2 + \frac{e\mathbf{A}_2}{c}), \qquad (5.4)$$

where $\mathbf{A}$ is the vector potential of the field $\mathbf{H}$ experienced by the corresponding electron (see Sec. 1.2.3).

In the ground state, i.e., at zero temperature and no field, $n = 0$ and $\widetilde{\mathbf{p}} = \mathbf{p}$. Hence, taking into account that $R \neq 0$ (otherwise, stable Cooper pairs could not exist), from Eq. (5.3) it follows

$$(\widetilde{\mathbf{p}}_{cp})_0 = (\mathbf{p}_{cp})_0 = \mathbf{p}_{10} + \mathbf{p}_{20} = 0, \qquad (5.5)$$

where an additional subscript 0 denotes zero field.

This is the equivalent of the Cooper condition Eq. (5.2) for the case of zero field and temperature.

One more important property of Cooper pairs at zero field is related to their ensemble: due to 3D symmetry of the medium in the absence of a field, the total magnetic moment of all pairs and the sum of instantaneous velocities of the paired electrons are zero[4], i.e.,

$$\mathbf{M}_0 = \sum \boldsymbol{\mu}_0 = 0, \qquad (5.6)$$

and

$$\sum \mathbf{v}_0 = 0. \qquad (5.7)$$

where summation is taken over all paired electrons in the specimen.

Eqs. (5.6) and (5.7) hold for the unit volume of the S phase, as well as for its physically infinitesimal element dV^5. In superconductors dV is a volume, which size is much smaller than the size of the volume taken by the S phase and much larger than the radius of electron orbit R_0, determining the scale of microscopic inhomogeneities of this phase.

However, Cooper's condition Eq. (5.2) is formulated without regard if the field is zero or not. What happens to $\widetilde{\mathbf{p}}_{cp}$ in the field?

When the applied field is turned on at zero temperature, the vector potential and the field intensity inside the specimen change from zero to $\mathbf{A}$ and

---

[4]Note a similarity with Eqs. (2.9) and (2.10) for a diamagnetic atom.

[5]As mentioned, by default we consider materials with homogeneous distribution of Cooper pairs at zero field.

H, respectively, over a short relaxation time. Each of the paired electrons experiences an action of the Lorenz force Eq. (2.11). In result, the kinetic linear momentum of one electron changes for $-e\mathbf{A}/c + \int \mathbf{F}_M dt$. Accordingly, $\tilde{\mathbf{p}}_{cp}$ in the field is

$$\tilde{\mathbf{p}}_{cp} = \tilde{\mathbf{p}}_1 + \tilde{\mathbf{p}}_2 = \left[\left(\mathbf{p}_{10} - \frac{e}{c}\mathbf{A}_1 + \int \mathbf{F}_{M1} dt\right) + \frac{e}{c}\mathbf{A}_1\right] +$$
$$\left[\left(\mathbf{p}_{20} - \frac{e}{c}\mathbf{A}_2 + \int \mathbf{F}_{M2} dt\right) + \frac{e}{c}\mathbf{A}_2\right] = \mathbf{p}_{10} + \mathbf{p}_{20} = 0, \quad (5.8)$$

where we took into account that $\mathbf{F}_{M1} = -\mathbf{F}_{M2}$ because the uniform field cannot move the pair's center of mass.

Thus, at zero temperature the generalized linear momentum of Cooper pairs is null regardless of the field. Therefore, the pairs are in the ground state all over the fields range of the S state at zero temperature.

But Cooper's condition is formulated without regard to temperature as well. Therefore n in Eq. (5.3) is zero, or Cooper pairs are in the ground state, over the entire range of both fields and temperatures of the S state. At the same time, the radius of the electron orbit R_0 changes with temperature[6] due to the changing number density of the pairs n_{cp}.

5.2.2   Meissner state

The Meissner state is defined as an equilibrium S state with $B = 0$ throughout the specimen volume V. Equivalently, this is the state with $\chi = -1/4\pi$ or $\mu_m = 0$ over V. Note that, as in ordinary diamagnetics, χ does not depend on either the field or temperature. For simplicity we will consider a specimen of cylindrical geometry[7]. The magnetic moment $\mathbf{M}$ and the magnetic energy E_m of such a specimen in the MS are

$$\mathbf{M} \equiv \chi V \mathbf{H} = -\frac{V}{4\pi}\mathbf{H}_0 \tag{5.9}$$

and

$$E_m \equiv -\int \mathbf{M} \cdot d\mathbf{H}_0 = -\frac{\mathbf{M} \cdot \mathbf{H}_0}{2} = \frac{V H_0^2}{8\pi}. \tag{5.10}$$

Upon turning on the field, microscopic charges in the specimen experience the action of Lorentz force $\mathbf{F}_L$ (Eq. (2.11)). In a result, closed microscopic currents formed by these charges, or corresponding microscopic magnetic moments, precess with the Larmor frequency.

---

[6]Below we will see that R_0 does not depend on the field due to the fulfillment of the Larmor condition.

[7]Due to the homogeneity of $\mathbf{H}$ in ellipsoidal bodies, the following consideration is applicable to a specimen of any geometry with the necessary corrections for the external field created by the magnetized specimen.

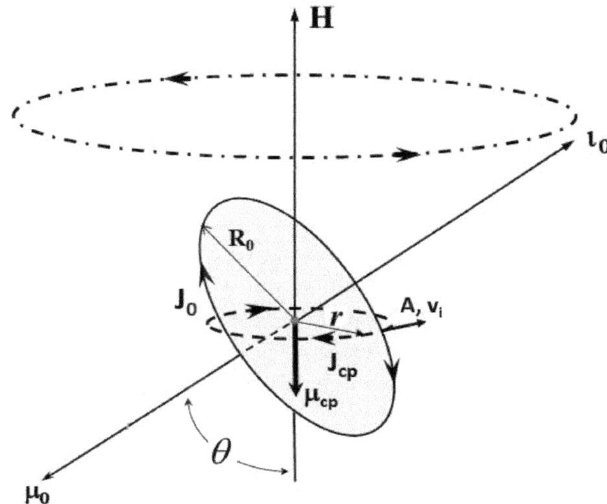

Figure 5.2   Schematics of the precessing Cooper pair in the field $\mathbf{H}$. Vectors ι_0 and μ_0, current J_0 and radius R_0 are the same quantities as those at zero field shown in Fig. 5.1. A dot-dashed circle designates the path of the tip of the precessing ι_0. A dashed horizontal circle depicts the field-induced current J_{cp}; it is also a line of the vector potential directed opposite to the current J_{cp}. $\mathbf{A}$ and $\mathbf{v}_i$ designate the vector potential and the induced velocity of one electron, respectively; r is the radius of the induced current; $\boldsymbol{\mu}_{cp}$ is the induced magnetic moment of the pair. Designated by a single arrow $\mathbf{A}$ and $\mathbf{v}_i$ are proportional but not equal. After Kozhevnikov [116]; reprinted with permission from Springer Nature.

The precession of all momenta, except those of Cooper pairs, leads to normal diamagnetic plus a possible paramagnetic responses, discussed in Ch. 2. Let us look what happens to Cooper pairs.

The forces $\mathbf{F}_M$ acting on each of the paired electrons, apart from being equal and opposite, are non-central. Therefore, they make a couple resulting, due to non-zero angular momentum ι_0, in precession of the pair. In turn, the precession leads to the appearance of an induced circular current J_{cp} in a plane transverse to $\mathbf{H}$, the magnetic moment of which $\boldsymbol{\mu}_{cp}$ is directed opposite $\mathbf{H}$. The precessing Cooper pair is schematically depicted in Fig. 5.2.

To calculate $\boldsymbol{\mu}_{cp}$, we just need to repeat the steps leading from Eq. (2.11) to (2.19), as it is done in [116]. In this case $\boldsymbol{\mu}_i$ in Eq. (2.19) is an average magnetic moment per one electron induced in each pair. The magnitude of $\boldsymbol{\mu}_{cp}(= 2\boldsymbol{\mu}_i)$ is

$$\mu_{cp} = \frac{\pi r_i^2}{c} J_{cp} = \frac{\pi r_i^2}{c} \frac{(2e)}{2\pi} \omega_i = \frac{(2e)^2 r_i^2}{4(2m)c^2} H = \frac{(r_i/2)^2}{\lambda_L^2} \frac{H}{4\pi n_{cp}}, \qquad (5.11)$$

where the induced current $J_{cp} = 2J_i$ (Eq. (2.18)), ω_i is the frequency of this current equal to the Larmor frequency (Eq. (2.17)), $r_i = \sqrt{\langle r^2 \rangle}$ is the rms radius[8] of the induced currents averaged over the pairs in the volume element dV defined above, n_{cp} is a number density of Cooper pairs equal to $n_s/2$, and λ_L is the London penetration depth.

In terms of parameters of Cooper pairs, λ_L is

$$\lambda_L \equiv \left(\frac{mc^2}{4\pi n_s e^2} \right)^{1/2} = \left(\frac{m_{cp}c^2}{4\pi n_{cp}q_{cp}^2} \right)^{1/2}, \tag{5.12}$$

where $m_{cp} = 2m$ and $q_{cp} = 2e$ are the mass and charge of the pair, respectively; m is the mass of a free electron, e is its charge.

The magnetic momenta induced in all pairs are parallel, so the magnetic moment of our specimen is[9]

$$\mathbf{M} = n_{cp}V\boldsymbol{\mu}_{cp} = -\frac{(r_i/2)^2}{\lambda_L^2}\frac{V}{4\pi}\mathbf{H} = -\frac{(r_i/2)^2}{\lambda_L^2}\frac{V}{4\pi}\mathbf{H}_0. \tag{5.13}$$

Comparing the latter with Eq. (5.9), we find

$$r_i = 2\lambda_L. \tag{5.14}$$

It is easy to see that this result is valid for specimens of any geometry.

Thus, similar to R_0, the rms radius of the induced current r_i is determined by the temperature-dependent pairs density n_{cp}. Therefore, r_i is proportional to R_0.

Before going further we should check the Larmor condition. Taking typical value of $H_{c1} \sim 100$ Oe and $\lambda_L \sim 10^{-6}$ cm, from Eq. (2.16) we find $v_i \sim 10^2$ cm/s, which is six orders of magnitude less than Fermi velocity $v_F \sim 10^8$ cm/s. So, Larmor's condition is fulfilled with a gargantuan margin. Therefore (see Sec. 2.1), parameters of Cooper pairs at zero field, such as R_0 and $|\boldsymbol{\mu}_0|$ stay unchanged in the field, and the rms radius of the induced motion r_i and, consequently, n_{cp}, do not depend on the field either. This explains the field-independence of χ in regular diamagnetics and superconductors in the MS.

The fulfillment of the Larmor condition also means that Eqs. (5.6) and (5.7) hold both in zero and non-zero field[10]. Consequently, ΔT, the change of kinetic energy of electrons in the specimen is equal to the sum of kinetic energies of their field-induced motion $\sum \epsilon_{cp}$. In turn, according to the energy conservation, in cylindrical specimens $\Delta T = E_m$ (see Sec. 2.3.2).

---

[8]The radiuses of induced current r are different in pairs with different orientation of $\boldsymbol{\mu}_i$ with respect to $\mathbf{H}$.

[9]The first two of these equalities held for specimens of any geometry; the last equality ($\mathbf{M}$ in terms of $\mathbf{H}_0$ is valid when $\eta = 0$.)

[10]Like the electron orbits in regular diamagnetics, all pairs precess synchronously, maintaining the mutual orientation of $\boldsymbol{\mu}_0$ and $\mathbf{v}_0$ which they had in the field absence.

An average kinetic energy of the induced motion of electrons in Cooper pairs, or an excess kinetic energy of the paired electrons acquired in the field, is

$$\epsilon_{cp} = 2\frac{mv_i^2}{2} = m\left(\frac{e}{2mc}Hr_i\right)^2 = \frac{H^2}{8\pi n_{cp}}, \tag{5.15}$$

where the expression in parentheses is $\mathbf{v}_i$ from Eq. (2.16) in which we took into account that $\mathbf{r} \perp \mathbf{H}$ and $r_i = 2\lambda_L$.

The total change of kinetic energy of electrons in our specimen is

$$\Delta T = \sum \epsilon_{cp} = \epsilon_{cp} n_{cp} V = \frac{H^2}{8\pi}V = \frac{H_0^2}{8\pi}V, \tag{5.16}$$

where the summation is taken over all Cooper pairs.

Comparing this expression with Eq. (5.9) we see that the MWM meets the law of energy conservation[11].

Next, we calculate the gyromagnetic ratio of the specimen. Using Eqs. (2.17) and (5.10) we write

$$\gamma \equiv \frac{M_i}{L_i} = \frac{\mu_{cp} n_{cp} V}{\iota_{cp} n_{cp} V} = \frac{\mu_{cp}}{(2m)v_i r_i} = \frac{e}{2mc}, \tag{5.17}$$

where M_i and L_i are magnitudes of the induced magnetic moment and of the angular momentum of the specimen, respectively; and ι_{cp} is an average induced angular momentum per Cooper pair.

We see that γ in MWM agrees with the experimental result of I. Kikoin and Goobar. At the same time, Eq. (5.17) confirms I. Kikoin's interpretation [62], that the gyromagnetic effect in superconductors is caused by microscopic currents in their bulk but not by the eddy current on the surface.

After all, the induction within the specimen is

$$\mathbf{B} = \mathbf{H} + 4\pi\mathbf{I} = \mathbf{H} + 4\pi(\boldsymbol{\mu}_{cp} n_{cp}) = \mathbf{H} - 4\pi\frac{\mathbf{H}}{4\pi n_{cp}}n_{cp} = 0, \tag{5.18}$$

where we used Eqs. (5.11) and (5.14), and the fact that $\boldsymbol{\mu}_{cp}$ is negative.

Strictly speaking, the way how magnetization is derived in Sec. 2.1 is applicable to superconductors cooled in zero field (ZFC). However, it is valid for FC specimens as well. Indeed, upon lowering temperature below $T_c(H_0)$ in the field $H_0 < H_{c1}$, a temperature dependent fraction of conduction electrons condenses forming stable Cooper pairs, which, being in the field, start precessing. Thus, the condensation in the FC specimen is equivalent to turning on the field in the ZFC one. In both cases, after a short-time relaxation, the environment inside the specimen, i.e., $\mathbf{A}$, $\mathbf{H}$, $\mathbf{I}$ and $\mathbf{B}$, is the same. Therefore, the MWM does indeed reproduce the Meissner effect[12].

---

[11]In the standard theories the description of the Meissner effect is based on the London theory in which this law is broken (see Eq. (3.27)).

[12]Recall that in the London theory zero induction is achieved by postulating (see Sec. 3.2).

Due to microscopic character of the induced currents, the Meissner condition ($B = 0$) can be established in specimens/domains of any shape, provided the field **H** is uniform. The latter is indeed so in ellipsoidal superconductors, regardless of whether they are in the homogeneous or inhomogeneous equilibrium states. This explains why the MS, IS and MXS are observed only in the ellipsoidal bodies, as well as the variety of (never ellipsoidal!) domain shapes in the IS (see, e.g., Figs. 4.1 and 4.7).

5.2.3 Microscopic whirls

Same as in regular diamagnetics, the net effect of the field in superconductors in the MS is the appearance of induced microscopic circular currents in the entire specimen, lying in planes perpendicular to **H**. The currents have identical strength $J_{cp} = 2J_i$ and the same rms radius r_i. How are these currents arranged with respect to each other?

The symmetry following from the uniformity of the field **H** suggests two options: either complete chaos or complete order. Since the free energy of the system of ordered currents is lesser, the second option prevails.

The identical circular currents in the uniform field possess the maximum symmetry if they form a 2D hexagonal lattice of cylindrical micro-whirls, which are similar to densely packed and tightly wound solenoids aligned with **H**. The whirls are infinite, since their length is macroscopic (equal to the specimen size along the direction of **H**) while the diameter is microscopic $(2r_i = 4\lambda_L)$[13].

Since the whirls/solenoids are parallel, they do not interact, in consistency with the fact that the magnetic energy of superconductors E_m is equal to the sum of kinetic energies of ϵ_{cp} (Eq. (5.14)), i.e., it does not contain terms responsible for interactions. Recall that the absence of interaction between the induced currents follows from the magnetization curves of the MXS discussed in Sec. 4.4.2. In addition, since the currents are formed by the field-induced circular motion of the paired electrons, the complete ordering implies that in all Cooper pairs this motion is in phase[14]. One can object, however, that currents of neighboring loops in the whirl must attract each other and therefore interact. This is true, but the net interaction between the loops vanishes due to longitudinal symmetry (the symmetry along direction of **H**). This means that the loops are equally spaced. Let us compute this spacing.

As we know, due to the continuity of the normal component of induction B_n, the applied field bends around the specimen in the MS as shown in Fig. 3.10. In its turn, the tangential component of induction B_t experiences

---

[13]Each whirl can be imagined as an endless solenoid of a tiny diameter $2r_i$ "wound" on a field line.

[14]Note the direct analogy with the phase of the quantum-mechanical wave function. The wave functions of all Cooper pairs must be in phase since otherwise there would be a total current forbidden by the time reversal symmetry.

a jump[15] at the surface caused by the field-induced surface currents formed by electrons bound in Cooper pairs. From the boundary condition for B_t (Eq. (1.25)), the linear density of the surface current in our cylindrical specimen is

$$g_b \equiv \frac{J}{L} = \frac{cH_0}{4\pi} = cI, \qquad (5.19)$$

where J is the surface current and L is the length of the specimen.

Since $I = \mu_{cp}n_{cp}$, the expression for g_b is

$$g_b = c\mu_{cp}n_{cp} = J_{cp}\pi r_i^2 n_{cp}\frac{V}{A_c L} = J_{cp}\frac{\pi r_i^2}{A_c}(n_{cp}V)\frac{1}{L}, \qquad (5.20)$$

where A_c is the cross-sectional area of the specimen, equal to V/L.

Let us denote the number of the current loops in A_c as $N_\perp = A_c/\pi r_i^2$ and the number of loops along L as $N_\parallel$. The total number of the loops is equal to the number of pairs $N_{cp} = n_{cp}V = N_\perp N_\parallel$. Then,

$$g_b = J_{cp}\left(\frac{N_{cp}}{N_\perp}\right)\frac{1}{L} = J_{cp}\frac{N_\parallel}{L}. \qquad (5.21)$$

The last fraction is the number of loops per unit length. Denoting $N_\parallel/L = n_\parallel$ and using Eqs. (2.18) and (5.19) we find that the spacing between the induced current loops Δ is

$$\Delta = \frac{1}{n_\parallel} = J_{cp}\frac{4\pi}{cH_0} = 2\frac{e^2 H_0}{4\pi mc}\frac{4\pi}{cH_0} = \frac{2e^2}{mc^2} =$$

$$5.6 \cdot 10^{-13}\,cm = 5.6\,fm. \qquad (5.22)$$

Thus, the loops are indeed very tightly "wound" and Δ is a universal number[16], about three times the size of a proton (1.7 fm). In terms of λ_L the spacing is

$$\Delta = \frac{1}{\lambda_L^2 n_s 2\pi} = \frac{1}{4\pi n_{cp}\lambda_L^2} = \frac{1}{\pi n_{cp}r_i^2}. \qquad (5.23)$$

Note that the constancy of Δ follows from the fact that $r_i^2 = 4\lambda_L^2 \sim 1/n_{cp}$[17]. The same fact provides a full coverage of the specimen cross-sectional area by the induced currents when temperature (and therefore n_{cp}) changes.

---

[15]The jump is a charge occurring over a distance, characterizing microscopic inhomogeneities of the medium in question. In normal materials this distance is of the order of radius of the field-induced circular currents. As a rule, this radius is of the order of atomic size (see Sec. 2.3.2). In superconductors this distance is of the order of r_i or R_0, which are proportional quantities of the same (or about the same) order.

[16]More precise value of Δ is $(2e^2/mc^2)\pi\sqrt{3}/6 = 5.1\,fm$.

[17]This statement can be read in opposite way: the reversed proportionality between λ_L^2 and n_{cp} follows from the longitudinal symmetry along the field $\mathbf{H}$, i.e., from the uniformity of $\mathbf{H}$.

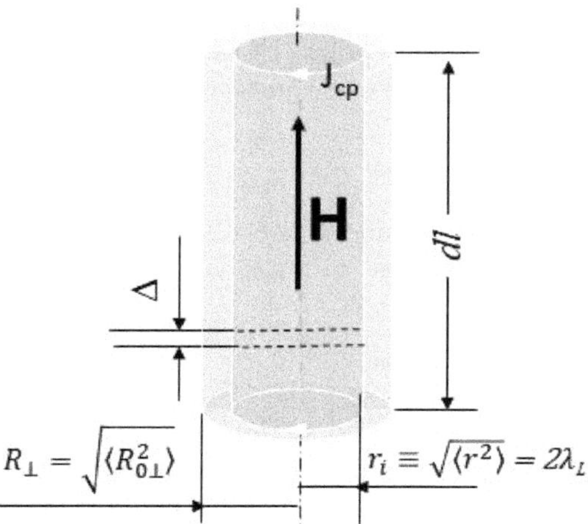

Figure 5.3 Diagram of an element of an individual micro-whirl. Its length $dl \gg R_0$. $R_\perp$ is the rms projection of the pairs radii R_0 onto the transverse (to **H**) plane averaged over all possible orientations of the electron orbits in zero field; the cylinder of this radius (shown in light gray) is filled by orbiting Cooper pairs, which magnetic moments μ_0 obey Eq. (5.6). r_i is the rms radius of the induced currents in the pairs J_{cp}, laying in the transverse planes; these currents form the solenoid-like cylinder shown in gray. $\triangle$ is the spacing between individual induced current loops. **H** is the field intensity within the specimen. Note that $R_\perp$ and r_i are averaged quantities, so the walls of the depicted double-wall cylinder are not sharp.

One more parameter characterizing an individual whirl is $R_\perp = \sqrt{\langle R_{0\perp}^2 \rangle}$, the rms projection of the pairs radii onto the plane transverse to **H**; referring to Eq. (2.30), $R_\perp = R_0 \sqrt{2/3}$. $R_\perp$ is the rms radius (averaged over all possible angles θ in Fig. 5.2) of a cylinder filled by Cooper pairs of all orientations which form the given whirl. As in normal diamagnetics, r_i can be greater or lesser than $R_\perp$. Respectively, each micro-whirl represents an infinite double-wall cylinder with radii $R_\perp$ and r_i. The whirl element and the whirl structure are schematically shown in Figs. 5.3 and 5.4, respectively. Important to emphasize that $R_\perp$ and r_i are averaged quantities, i.e., walls of the whirls are not sharp, which also follows from the uncertainty principle.

Finally, we introduce the parameter $\aleph$ (aleph), defined as

$$\aleph = \frac{r_i}{R_\perp}. \tag{5.24}$$

From the above it follows that $\aleph$ is a constant depending neither on the field nor on temperature. Below we will see that $\aleph$ is a material constant close in its essence to the GL parameter κ but defined for the entire temperature/field range of the S state.

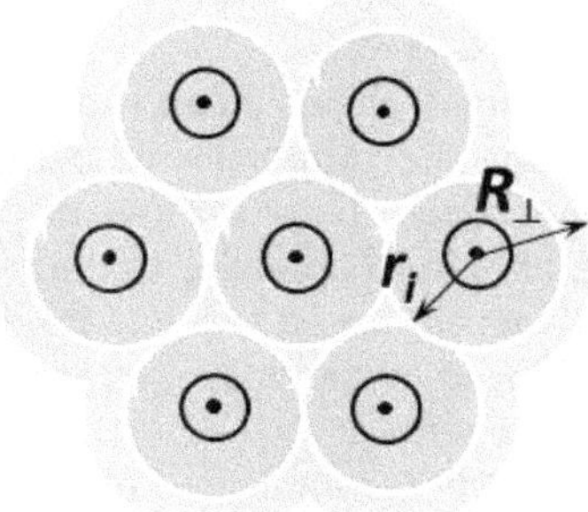

Figure 5.4 The micro-whirls structure in the plane transverse to **H**. Dots in small circles designate tips of the vector **H**; white contours with arrows are the induced currents J_{cp} "wound" on the H-lines. See Fig. 5.3 for other notations.

5.2.4 Entropy, temperature and resistivity

The complete ordering of the field-induced currents together with Eq. (5.6) means that entropy of the ensemble of Cooper pairs S_{cp} is zero[18]. As we saw in Sec. 4.3.2, zero entropy of the ensemble of Cooper pairs also follows from thermodynamic reasoning based on the experimental fact of the temperature independence of χ. Recall that zero entropy was for the first time deduced from the absence of thermoelectric effects in superconductors, discussed in Sec. 3.1.2. Therefore the MWM is consistent with all these facts and justifies the zero-entropy postulate of the two-fluid model of Gorter and Casimir.

The complete ordering (i.e., zero entropy) leads to one more very important property of the ensemble of Cooper pairs. As follows from the Third law, temperature of a statistical system with zero entropy is zero. Hence, temperature of the ensemble of Cooper pairs T_{cp} is zero, regardless on the specimen temperature T. This fact is consistent with (or essentially follows from) Cooper's condition Eq. (5.2) and the Bohr-Sommerfeld quantization condition Eq. (5.3) for the ground state, according to which the center of mass of the paired electrons is motionless. At the same time, each of the paired electrons continues to move with the Fermi speed ($\sim 10^8$ cm/s) which it had before the condensation. Recall that temperature of the unpaired conduction electrons equals the Fermi temperature $T_F \sim 10^4$ K, which is about the same as the Sun surface temperature.

The truly stunning transformation of the "very hot" single electrons to the "deadly cold" electron pairs can be compared with the fact known from

---

[18]This can also be seen from the fact that number of accessible states for the induced magnetic moments is unity in all diamagnetics, including superconductors. Accordingly, $S_{cp} = k_B \ln 1 = 0$, where k_B is the Boltzmann constant.

the relativity theory: two flying apart massless photons form a massive pair located in their center of mass [171].

However, $T_{cp} = 0$ certainly does not mean that the specimen (ionic lattice) temperature is irrelevant for the properties of paired electrons, since otherwise T_c would be infinite. Indeed, pairs are not in free space, they owe their existence to the lattice. Changing T changes the lattice polarization, resulting in a change of n_{cp}, as it was shown in the two-fluid model (Eq. (3.10)). But at all temperatures the induced currents remain strictly ordered, thus keeping S_{cp} and, hence, T_{cp} equal to zero.

One of the most remarkable consequences of $T_{cp} = 0$ is the zero resistance experienced by total electrical current in superconductors. Let us see how it comes about.

As was first recognized by Shoenberg (see Sec. 3.1.4) there are two kinds of dissipation-free currents in superconductors: induced microscopic circular currents responsible for magnetization and total currents running on macroscopic paths. The latter includes transport current in hybrid S/N circuits powered by an external e.m.f., and the persistent current encircling openings with trapped flux in multiply connected bodies powered by an electromagnet[19].

The microscopic currents are similar to the molecular currents in conventional diamagnetics. In both cases these are currents caused by the orbital motion of electrons bound either in atoms or in Cooper pairs. In a magnetic field these currents precess creating microscopic diamagnetic moments in accord with the Larmor theorem. A colossal quantitative difference in susceptibilities of the normal and superconducting diamagnetics (up to 100 000 times!) is due to the difference in the size of the orbits, i.e., in the radius r_i of the induced current in atoms and Cooper pairs.

However, there is also an important qualitative difference. In conventional diamagnetics the orbiting electrons are bound in motionless atoms (atoms have no translational degrees of freedom), while in superconductors - in movable Cooper pairs. It is the very mobility of the pairs that allows them to organize into an ideally structured whirl lattice. On the other hand, since the pairs have electric charge, they can form the total current.

Zero temperature T_{cp} implies that the pairs in the micro-whirls do not communicate/interact with the environment, i.e., with their neighbors. Therefore, the ordered pairs behave like free charge carries in the medium in question. In the total current the ordered micro-whirls move with a drift velocity v_d relative to the ionic lattice, experiencing no resistance until the current reaches a critical value, considered in the next section. Therefore, since magnetic field $\mathbf{H}$ is always present in any body[20], the total current in superconductors is carried by electrons which motion represents a combination of the rotational motion in the whirls and translational motion along with the whirls.

Note a close similarity of the total current in superconductors to the superfluid current in He-II, where the latter also represents a combine

---

[19]The electromagnet does work to set up this current.

[20]This field is caused by an external field (which can never be completely compensated) and, in the case of the total current, by this current itself.

circular/translational motion[21] [172, 173]. Such motion of paired electrons agrees and explains observations of Meissner and Ochsenfeld in their fourth arrangement[22].

One more important aspect of the fact that $T_{cp}(= 0)$ is independent on the specimen temperature T, is the absence of a fundamental limitation on the critical temperature T_c, as was demonstrated in research on high-temperature superconductors following the discovery of Müller and Bednorz [9].

Summarizing the last three sections, we note that the MWM meets the necessary requirement of a superconductivity model: it reproduces the three big zeroes, i.e., zero resistance, zero induction and zero entropy. All of them have a single root: the generalized Bohr-Sommerfeld quantization condition for the paired electrons with antiparallel spins.

5.2.5 Critical current and Silsbee's hypothesis

The destruction of superconductivity by electric current was discovered by Kamerlingh Onnes in 1913, when, in attempt to measure the residual resistance of mercury wire below T_c, he found a sudden reappearance of resistance at an increased current [174]. Soon the same effect was revealed in lead and tin [175]. The current J_c that destroys superconductivity is called the critical current. The value of J_c decreases with increasing temperature, approaching zero near T_c; an external magnetic field, including the field produced by the current itself in a superconducting wire wound in a coil, also decreases J_c.

Based on Onnes's data on J_c and the critical field H_c (discovered in 1914 [176]) in 1916 Francis Silsbee put forward a hypothesis stating that "The threshold value of the current is that at which the magnetic field due to the current itself is equal to the critical magnetic field" [177].

Specifically, a current J flowing in a cylindrical wire of radius a produces a magnetic field around the wire, the strength of which at the wire surface is

$$H = \frac{2J}{ca}. \tag{5.25}$$

Then, according to Silsbee's hypothesis,

$$J_c = \frac{ca}{2} H_c. \tag{5.26}$$

A problem with this hypothesis is that the critical current density $j_{cr} = J_c/\pi a^2 \sim 1/a$, decreases with the wire radius, whereas H_c is the intrinsic

---

[21]At the same time, liquid He-II becomes superfluid only below the lambda-point (2.17 K), while above it the liquid helium is a normal fluid. On the contrary, the ensemble of Cooper pairs superconducts at any $T < T_c$, in accord with the fact that $T_{cp}(= 0)$ is independent of the specimen temperature.

[22]Quoting Meissner and Ochsenfeld [7], "When the parallel superconductors are connected end-to-end in series and an external current is connected to flow through them above the critical temperature the magnetic field between the superconductors is increased below the transition temperature the external current being unchanged".

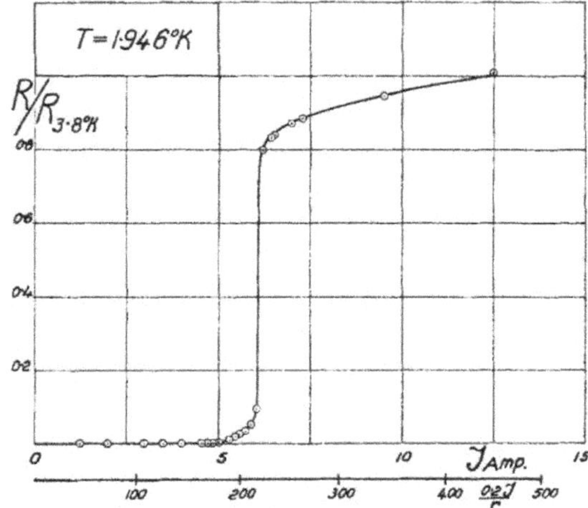

Figure 5.5 Transition curve for the destruction of superconductivity in tin cylindrical wire by an electric current. $R/R_{3.8K}$ is the wire resistance in units of its resistance slightly above $T_c(= 3.72K)$. The second horizontal axis is the field caused by the current on the wire surface in G. The measurements were conducted at constant temperature 1.946 K. After L. Shubnikov and N. Alexeyevski [179, 180]; reprinted from [180].

parameter of material. For this reason, Silsbee's hypothesis, deserving of the name Silsbee's paradox, has attracted considerable attention.

To test it, Tuyn and Onnes measured resistance vs current of (a) straight superconducting tin wires of different diameters and (b) a cylindrical tin film deposited on a glass tube when an additional current was flowing in a copper wire stretched along the tube axis; both currents could be tuned independently. In this case the field on the surface of the tubular sample is equal to the algebraic sum of the fields due to both currents. The results obtained were consistent with Silsbee's hypothesis [178][23].

Shubnikov and Alexeyevski [179] were the first to measure the entire $R - J$ (resistance R vs current J) curve in the transition region[24]. A detailed report

---

[23]This paper was published in April 1926 with an editorial preface "While the manuscript for this paper was on the ocean in transit from Leiden to Philadelphia, news of the death of Dr. H. Kamerlingh Onnes appeared in the daily papers. His death takes a brilliantly successful figure out of the world of experimental science, and deprives the realm of scholarship of an important contributor". Heike Kamerlingh Onnes passed away on February 21, 1926.

[24]The critical current density at the sudden change of resistivity was so enormous $(1.5 \times 10^5 A/cm^2)$ that it was a huge challenge to keep constant temperature at crossing J_c with increasing current. To solve this problem Shubnikov and Alexeyevski conducted their measurements at temperature below the helium lambda point (2.17 K). Note that superfluidity was yet to be discovered, but the authors already knew about an extremely high thermal conductivity of liquid helium below the λ-point from their own studies. This work was the first practical application of superfluidity.

was published by Alexeyevski [180][25]. The obtained $R - J$ curve for a poly-crystalline tin wire at zero applied field is shown in Fig. 5.4.

On one hand, these results perfectly confirmed Silsbee's hypothesis[26], but on the other, they brought in a new puzzle. It turned out that at $J = J_c$ not all superconductivity is destroyed: the resistance R jumps up not to its value in the N state R_N, but to about $0.8R_N$. With a further current increase, superconductivity exhibits a "plume", i.e., R/R_N gradually approaches unity. In a more detailed study of Scott [181] the plume appeared to be longer[27] than that in the experiment of Shubnikov and Alexeyevski. At the same time, Scott confirmed the magnitude of the jump $\approx 0.8R_N$.

The paradoxity of the plume lies in the fact that, based on the symmetry of the cylindrical sample with current, at $J > J_c$ superconductivity should be concentrated near its axis. However, if this is the case, then superconductivity should form a resistance-less filament, which should short-circuit the potential leads and, therefore, the resistance of the sample should be zero!

For alloys (type-II superconductors with H_{c2} much greater than H_c in type-I materials), it was found that Silsbee's hypothesis is valid if the field on the surface is less than H_{c1}, but it does not work at all if the field is above this value[28]. A conclusion was that Silsbee's hypothesis is inapplicable to type-II superconductors above H_{c1} [182]. As we already know (see Sec. 3.1.1) the transition to the resistive state in these materials above H_{c1} occurs due to the motion of flux lines and by no means related to H_{c2}.

A few theoretical scenarios for the plume of superconductivity were proposed. The most frequently cited one was suggested by F. London [35]. He assumed that at all currents above J_c a cylindrical sample is in the IS and H inside it has only one, angular, component in a cylindrical coordinate system (r, θ, z), herewith this component $H_\theta = H_c = const$ regardless on r. According to the next assumption, the S-fraction forms a chain of domains of a rhomboidal cross section, which *connect* with each other in the rhomboids vertices along the sample axis (see, e.g., [85] for a diagram). London was able to reproduce qualitatively the experimental $R - J$ curve, however with significant quantitative differences[29]. Admitting this fact, the author attributed it to the difficulties of the experiment.

Without going into detail, we note that F. London's scenario is unrealistic due to the contradictory assumptions underlying it. In particular, the

---

[25] After Shubnikov's arrest in 1937, his name was taboo in the USSR until 1957.

[26] The measured J_c corresponded to a field on the surface of the sample (tin wire with radius 0.0056 cm) equal to 219 G; the independently measured H_c was 218 G; the error bar was a few percent.

[27] Scott measured the $R - J$ curves for indium wires at decreasing current starting from $J \approx 2.4J_c$. At this maximum current R/R_N was still slightly less than unity.

[28] For example, at 4.2 K the field on the surface of Pb-Bi wire at the critical current was 308 G, whereas $H_{c2} = 16000$ G [182].

[29] According to F. London, the resistance jump is strictly $0.5R_N$ (which follows from the demagnetizing factor of the wire $\eta = 0.5$), and R/R_N approaches unity asymptotically, meaning that the plume is infinite.

assumption $H_\theta = H_c = const$ means that vectors $\mathbf{H}$ inside the sample are not parallel and therefore the field of these vectors is not uniform. Indeed, the uniformity of $\mathbf{H}$ in a cylindrical current-carrying sample conflicts with the sample symmetry and is therefore impossible. But the uniformity of $\mathbf{H}$ is the necessary condition for the IS, as well as for other equilibrium states (see Ch. 4). Hence, the sample in question can never be in either of these states. The reader can easily notice other inconsistencies in F. London's interpretation.

Using the MWM, Silsbee's hypothesis can be explained as follows.

By definition of the transport current[30], the current in the S section of a hybrid N/S circuit represents the collective motion of Cooper pairs. In other words, in the circuit with current J the ensemble of ordered pairs[31] is moving as a whole with the drift velocity v_d. Then, according to Ampere's law, the magnitude of the field intensity H in the cross section of a cylindrical sample of radius a carrying the current J is

$$H = \frac{2J}{ca^2}r, \tag{5.27}$$

where $r(\leq a)$ is the distance from the sample axis or the radius of concentric lines of the field intensity inside the sample.

Due to this current-induced field, the pairs precess and form micro-whirls. In the given case the latter represent solenoids of microscopic radius r_i, wound around the field lines of radius r. Except near the axis, r is macroscopic. So at $J < J_c$ for most whirls the curvature of the field lines can be neglected. A schematic diagram of the whirl structure is shown in Fig. 5.6.

Due to precession, each paired electron acquires the additional velocity $\mathbf{v}_i$ and, accordingly, an excess kinetic energy ϵ_{cp}. If $v_i \gg v_d$, then the kinetic energy due to the drift velocity is negligible, and ϵ_{cp} is equal to (see Eq. (5.15))

$$\epsilon_{cp}(r) = \frac{(2m)v_i^2}{2} = \frac{H^2(r)}{8\pi n_{cp}}. \tag{5.28}$$

Applying Eq. (5.27), we find that near the surface (at $r = a$) this energy is

$$\epsilon_{cp}(a) = \frac{H^2(a)}{8\pi n_{cp}} = \frac{1}{8\pi n_{cp}}\left(\frac{2J}{ca}\right)^2. \tag{5.29}$$

When $\epsilon_{cp}(a)$ reaches the pair bonding energy e_{cp} defined in Eq. (4.6), the pairs near the surface break up. This leads to decrease in a, which becomes the radius of a superconducting core carrying the *entire* current[32]. Hence,

---

[30]The transport electric current is defined as a collective motion of charge carriers with a drift velocity $\mathbf{v}_d$; then from the charge conservation, the current density $\mathbf{j}$ is equal to $qn\mathbf{v}_d$, where q and n are the charge and number density of the carriers, respectively.

[31]If an external field is absent (i.e., negligibly small), the ordering of Cooper pairs occurs due to the field created by the collective motion of the pairs themselves.

[32]When the core is superconducting (i.e., resistanceless), it short-circuits the ends of the sample, thereby excluding the contribution of normal electrons to the total current.

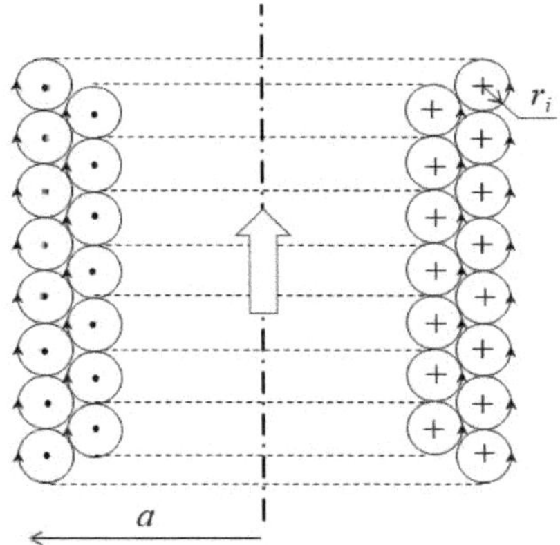

Figure 5.6 The micro-whirls structure in a cylindrical wire carrying the current $J \leq J_c$. a is the wire diameter; $r_i (\ll a)$ is the rms radius of the induced current in Cooper pairs equal to the whirl radius; small arrows indicate the direction of induced current J_{cp}; the fat arrow indicates the direction of the current J; dots and crosses are tips and tails of arrows indicating direction of the circular lines of the field **H** caused by the current J.

H increases to H_c at the core surface destroying the pairs there, and so on. Thus we have an avalanche-like collapse of superconductivity at the current J_c, which according to Eq. (5.29) is

$$J_c = \frac{caH_c}{2}. \tag{5.30}$$

During this collapse (specifically, the collapse of the superconducting core) the current density j equals its critical value j_{cr} only near the core surface. Denoting the critical value of v_i as v_{ic} (v_i at which $\epsilon_{cp} = e_{cp}$), from Eq. (5.29) one finds (see problem 5.1)

$$j_{cr} = n_{cp}(2e)v_{ic} = \frac{cH_c}{2\pi r_i}, \tag{5.31}$$

where r_i, the radius of the induced currents, does not depend of the field.

Eq. (5.30) is Silsbee's formula Eq. (5.26). We see, that J_c depends on the wire radius a, in accord with experiments. However, the critical current density does not: like H_c, it is a constant of material depending on temperature. All this means that j_{cr} is much greater than the average density of the critical current $J_c/\pi a^2$. The latter immediately follows from an expression for j_{cr} in terms of J_c (use Eq. (5.30)), which is

$$j_{cr} = \frac{J_c}{\pi a r_i} = \left(\frac{J_c}{\pi a^2}\right)\frac{a}{r_i}. \tag{5.32}$$

For example, in the experiment of Shubnikov and Alexeyevski $J_c/\pi a^2 = 1.5 \cdot 10^5$ A/cm^2 and the wire radius $a = 0.0056$ cm. Taking $r_i \approx 100$ nm (such an estimate can be expected based on the LE-μSR data for indium [130, 116]), one finds $j_{cr} \sim 1 \cdot 10^8$ A/cm^2, i.e., it is three orders of magnitude greater than $J_c/\pi a^2$. Note, that the same estimate follows from the London theory [86][33].

One more important point about the critical current is well seen from Fig. 5.6: this is a bound current induced in practically stationary Cooper pairs (since $v_i \gg v_d$). Similar, although much weaker bound currents occur in regular diamagnetics with the transport current [19].

In practice, the destruction of superconductivity at increasing current is accompanied with the sudden release of an enormous amount of heat, often melting the sample[34]. However, if the temperature is maintained constant, superconductivity cannot be restored at $J \geqslant J_c$. Indeed, in this case the current is a flow of now normal electrons (with minor contribution of the pairs near the axis considered below) which creates the field following the same Eq. (5.27), as in the S state, but now the field at the surface is greater than H_c. Therefore, any attempt to restore superconductivity (i.e., the superconducting core of any macroscopic radius) would instantly fail.

However, as we just mentioned, after collapse of superconductivity at $J = J_c$, a minor fraction of Cooper pairs (of an order of r_i/a form the total amount at $J < J_c$) "survived" in proximity of the sample axis. These survived Cooper pairs are responsible for the plume. The immediately arising question about the short-circuiting superconducting filament can be answered as follows.

In order to have zero resistance, the filament should be oriented along the field, as in Fig. 5.3. However, in the given case the field (albeit small) is perpendicular to the filament. Therefore, in view of the tiny filament thickness ($\sim r_i$), Cooper pairs cannot create a well structured whirl lattice and, consequently, they are no longer free charge carriers. At the same time, being spinless, the paired electrons interact with the neighbors weaker than their normal fellows, and therefore total sample resistance is still less than that of the sample cleared of Cooper pairs.

At $J > J_c$ the condition $v_i \gg v_d$ is no longer fulfilled as well. This means that a contribution of the drift velocity $\mathbf{v}_d$ in ϵ_{cp} must be taken into account in Eq. (5.28). The last of survived Coopers pairs are located on the axis, where the field strength and, therefore, v_i are negligibly small (compared to H_c and v_d, respectively). Thus, the condition for destroying these pairs, i.e, for

---

[33]In the London theory (see Sec. 3.2.2), the critical current is the total circumferential (screening) current running near the surface of a cylindrical specimen in the MS at $\mathbf{H}_0 = \mathbf{H}_c$. Assuming that this current runs within the penetration depth λ_L, its density is $j_c = cH/4\pi\lambda_L$. Formally, this is the same expression as Eq. (5.31), but with a cardinally different physical meaning.

[34]Such a "superconducting quench" for the first time was observed by Onnes when he discovered the critical current. In 19th September of 2008 the quench incident happened while testing the Large Hadron Collider causing extensive damage to over 50 superconducting magnets and infrastructure.

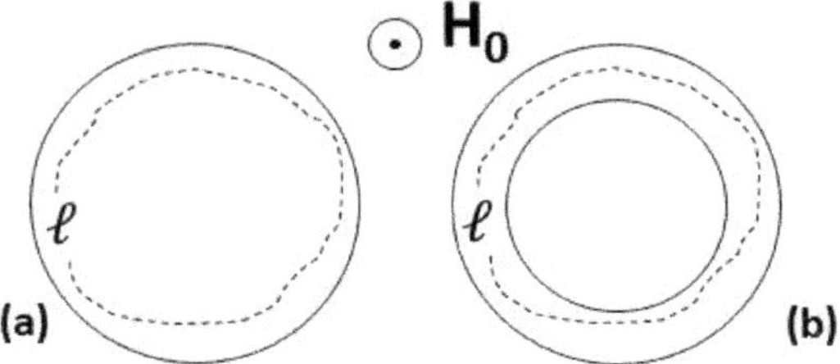

Figure 5.7 Cross section of a solid (a) and hollow (b) cylindrical superconductors in the field H_0 directed to the reader. The loop l lies away from the wall(s) of the cylinders.

$R/R_N = 1$, is

$$\epsilon_{cp} = \frac{2mv_{dc}^2}{2} = e_{cp} = \frac{H_c^2}{8\pi n_{cp}}, \tag{5.33}$$

where v_{dc} is the critical *drift velocity*.

Note the difference between this equation and Eq. (5.28). In the latter the bonding energy e_{cp} is reached in the orbital motion of the paired electrons near the wire surface, when the entire current is transported by Cooper pairs; whereas in Eq. (5.33) - in the translational one, when all remaining pairs move with the drift speed $v_{dc} = v_{ic}$ and practically all charge carriers are normal electrons. Denoting the current in this point as J_{c2}, one finds that its density j_{c2} is

$$j_{c2} = nev_{dc} = nev_{ic} = \frac{n}{2n_{cp}} j_{cr}, \tag{5.34}$$

where n is the density of normal electrons in the sample without Cooper pairs. One can easily figure, that $J_{c2}(= j_{c2}\pi a^2)$ is really huge ($\sim 10^4 J_c$)[35]. Hence, the curve R/R_N vs J approaches its upper limit nearly asymptotically. However, it does reach this limit at the very large, but still finite J_{c2}.

5.2.6 Flux quantization

Let us consider a specimen of cylindrical geometry in the MS and use the quantization condition Eq. (5.3) to calculate the circulation of $\widetilde{\mathbf{p}}_{cp}$ for an arbitrary *macroscopic* closed loop l. The loop lies, for simplicity, in the plane transverse to $\mathbf{H}$, as shown in Fig. 5.7a.

Moving along this loop, we will pass through lots of pairs, so we should consider the average generalized linear momentum $\langle \widetilde{\mathbf{p}}_{cp} \rangle$ and the average quantum number $\langle n \rangle$. Since all Cooper pairs are in identical conditions ($\mathbf{H}$ is uniform

---

[35]Normally, about 10% of the conduction electrons condenses into Cooper pairs [15].

throughout the specimen) $\langle n \rangle = n$. Therefore Eq. (5.3) takes the form

$$\oint_l \langle \widetilde{\mathbf{p}}_{cp} \rangle \cdot d\mathbf{l} = nh, \qquad (5.35)$$

where l is the loop length.

Now, let us open $\langle \widetilde{\mathbf{p}}_{cp} \rangle$ and take into account that in the MS $n = 0$. Then, Eq. (5.35) becomes

$$\oint_l \langle \widetilde{\mathbf{p}}_{cp} \rangle \cdot d\mathbf{l} = \oint_l \langle m\mathbf{v}_{i1} \rangle \cdot d\mathbf{l} + \oint_l \langle m\mathbf{v}_{i2} \rangle \cdot d\mathbf{l} + \frac{2e}{c} \oint_l \langle \mathbf{A} \rangle \cdot d\mathbf{l} = 0. \qquad (5.36)$$

The first two integrals are zero due to mutual compensation of the induced kinetic liner momentums of electrons in neighboring pairs (see Fig. 5.4).

To calculate the last integral we apply Stokes' theorem and write

$$\frac{2e}{c} \oint_l \langle \mathbf{A} \rangle \cdot d\mathbf{l} = \frac{2e}{c} \int_{F_l} (\nabla \times \langle \mathbf{A} \rangle) \cdot d\mathbf{f} = 0, \qquad (5.37)$$

where F_l is the area of a surface bounded by the loop l and $d\mathbf{f}$ is a vector element of this surface.

The integral over the area F_l is the flux of a vector $\nabla \times \langle \mathbf{A} \rangle$ and this flux equals zero. Therefore, since $\nabla \times \mathbf{A} \equiv \mathbf{H} \neq 0$, $\langle \mathbf{A} \rangle \neq \mathbf{A}$.

Inside our specimen the induction $\mathbf{B}$ and therefore its flux is zero (Eq. (5.18)). Therefore, Eq. (5.37) suggests that $\langle \mathbf{A} \rangle$ is the vector potential of the magnetic flux density $\mathbf{A}_B$ defined as $\mathbf{B} = \nabla \times \mathbf{A}_B$. In other words, the vector potential of the induction $\mathbf{A}_B$ is a macroscopic average of the vector potential $\mathbf{A}$ determining the field induced microscopic currents J_{cp}[36]. So, putting $\langle \mathbf{A} \rangle = \mathbf{A}_B$, we rewrite Eq. (5.37) as

$$\frac{2e}{c} \int_{F_l} (\nabla \times \langle \mathbf{A} \rangle) \cdot d\mathbf{f} = \frac{2e}{c} \int_{F_l} \mathbf{B} \cdot d\mathbf{f} = \frac{2e}{c} \Phi = 0, \qquad (5.38)$$

where Φ is the magnetic flux through the area F_l.

Now, let us take a tube-like hollow thick-wall long cylinder, apply the field $H_0 (< H_{c1})$ as shown in Fig. 5.7b and cool the cylinder below T_c. Then, consider the closed loop l encircling the cylinder's opening and laying inside the wall far (compare to r_i) from the inner and outer surfaces of the cylinder. The induction inside the wall is zero, implying that, as in Eq. (5.36), $\langle m\mathbf{v}_i \rangle = 0$. On the other hand, the flux Φ inside our hollow cylinder is frozen (see Sec. 3.1.4) and therefore it is *not zero*. Then, the circulation of $\langle \widetilde{\mathbf{p}}_{cp} \rangle$ is

$$\oint_l \langle \widetilde{\mathbf{p}}_{cp} \rangle \cdot d\mathbf{l} = \frac{2e}{c} \oint_l \langle \mathbf{A} \rangle \cdot d\mathbf{l} = \frac{2e}{c} \int_{F_l} \nabla \times \langle \mathbf{A} \rangle \cdot d\mathbf{f} = \frac{2e}{c} \int_{F_l} \mathbf{B} \cdot d\mathbf{f} = \frac{2e}{c} \Phi = nh.$$

$$(5.39)$$

---

[36] Note the exact match with the classical definition of $\mathbf{A}_B$ as a macroscopic mean of the vector potential [18].

Hence, the magnetic flux passing through the opening in a multiply connected superconductor is

$$\Phi = \frac{c}{2e}nh = \frac{\pi c\hbar}{e}n, \tag{5.40}$$

where $\hbar$ is the reduced Planck constant equal to $h/2\pi$.

This is the London's flux quantization but for the paired electrons. Hence, the origin of the superconducting flux quantization is the generalized Bohr-Sommerfeld quantization condition Eq. (5.3), as it should be.

Eq. (5.40) indicates that the superconducting flux quantum and therefore the flux passing through each flux line in type-II superconductors in the MXS (see Sec. 4.4.) is

$$\Phi_0 = \frac{c}{2e}h = \frac{\pi c\hbar}{e}. \tag{5.41}$$

As well known, the flux quantization and the single flux quantum in the flux lines are in full agreement with experiments [12, 13, 161, 160].

Thus, inside the S phase the vector potential $\mathbf{A}_B = \langle \mathbf{A} \rangle$, where $\mathbf{A}$ is the vector potential of the field $\mathbf{H}$. Now, what is the value of $\mathbf{A}_B$ in this phase?

In the plane perpendicular to $\mathbf{H}$ we have uniformly distributed identical induced circular currents J_{cp}, as shown in Fig. 5.4. Therefore, equal amount of electricity flows in opposite directions throughout an out-of-plane cross section of an arbitrary chosen volume element dV. Consequently, an average density of induced currents $\langle \mathbf{j}_i \rangle = en_s\langle \mathbf{v}_i \rangle$ is zero in any volume element of the specimen in the MS or inside the S phase. Hence, taking into account Eq. (2.15) we write

$$\mathbf{A}_B \equiv \langle \mathbf{A} \rangle = \langle \mathbf{v}_i \rangle = \langle \mathbf{j}_i \rangle = 0. \tag{5.42}$$

Thus, in the MWM the assumptions of the London theory concerning $\mathbf{j}_\infty$, $\mathbf{v}_\infty$ and $\mathbf{A}_\infty$ follow from the quantization condition Eq. (5.3) providing all of them are average of the corresponding microscopic quantities. At the same time, the unphysical assumptions about identity of superconductivity and vacuum (Eq. (3.20)) and the screening surface current in the MS are ruled out.

It remains to note that in the MWM the flux passes through the specimen when superconductivity ceases in at list one micro-whirl (see problem 5.2). Accordingly, when superconductivity disappears in one whirl, the flux passing through an appeared hole equals Φ_0, and the total flux passing through specimens with inhomogeneous magnetization (i.e., in the IS or MXS) is equal to the integer of Φ_0. Hence, the flux quantization is a consequence of the ordered whirl structure of the induced currents, which in turn follows from the generalized Bohr-Sommerfeld quantization condition Eq. (5.3).

Then, in full consistency with thermodynamics, in type-I superconductors in the IS, due to positive surface tension, the flux passes through multi flux-quantum N domains. In type-II materials, in which the surface tension is negative, the N domains represent the flux lines with the flux quantum enclosed in each of them.

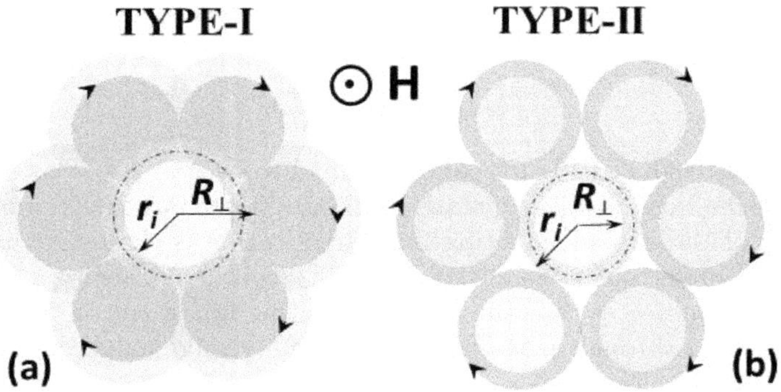

Figure 5.8 Schematics of the micro-whirls structure with the flux quantum Φ_0 passing through a type-I (a) and type-II (b) superconductors. In (a) $r_i < R_\perp$ or $\aleph < 1$; in (b) $r_i > R_\perp$ or $\aleph > 1$. The field **H** is directed toward the reader. The cylinders of radius $R_\perp$ (shown in light gray) are formed by Cooper pairs of all orientations constituting a given micro-whirl; the cylinders of radius r_i (shown in dark gray) are the micro-whirls formed by the induced currents J_{cp} shown with arrows; all these currents are in phase. Dash-dotted circuit in (a) and (b) is a notional S/N interphase boundary of the passing flux line. After Kozhevnikov [116]; reprinted with permission from Springer Nature.)

5.2.7   Surface tension

Properties of superconductors in states other than the MS are largely controlled by the S/N interface energy or the surface tension (see Ch. 4). As was shown by Pippard [139], to construct such an energy a theoretical model should possess two microscopic parameters with dimension of length. In MWM, such parameters are R_0, the orbital radius of the bound electrons, and r_i, the radius of the field-induced currents. Let us figure out how this fact leads to the surface tension of different signs.

In Eq. (5.24) we introduced the parameter $\aleph$. The only what we know so far about $\aleph$ is that it is positive. On the other hand, one can expect that properties of superconductors with $\aleph$ greater and lesser than unity are different. Below we will see that this indeed so.

For simplicity, we will again consider the cylindrical specimen. Cross sections of the current structure in specimens with $\aleph < 1$ and $\aleph > 1$ are schematically shown in Fig. 5.8. As in Fig. 5.3, the areas with radius $R_\perp$ are cross sections of the cylinders filled with Cooper pairs forming each micro-whirl. And those with radius r_i are cross-sections of the micro-whirls. We remind that both r_i and $R_\perp$ are root mean square radii, i.e. the edges of the cylinders are not sharp, they partially overlap not leaving voids.

The field passes through the specimen via a hole in the network of the micro-whirls. An elementary hole (the one carrying the single flux quantum Φ_0) appears at vanishing of one micro-whirl. To create such a hole superconductivity must cease in the corresponding cylinder with radius $R_\perp$. By definition, the magnitude of the minimal field H when it happens is H_{c1}.

Now, referring to Fig. 5.8a, the magnetic energy of the elementary hole Θ_n, where the subscript n stands for "normal", at $H = H_{c1}$ is

$$\Theta_n = \pi R_\perp^2 L \frac{H_{c1}^2}{8\pi} = \pi r_i^2 L \frac{H_{c1}^2}{8\pi} + \pi (R_\perp - r_i)(R_\perp + r_i) L \frac{H_{c1}^2}{8\pi}. \tag{5.43}$$

On the other hand, in the absence of a hole its space is taken by the cylinder, which magnetic energy Θ_s (the subscript s stands for "superconducting") at $H = H_{c1}$ is

$$\Theta_s = \pi r_i^2 L \frac{H_{c1}^2}{8\pi}. \tag{5.44}$$

Therefore, the difference Γ of the magnetic energies of the non-superconducting hole and the superconducting "insert" into this hole is

$$\Gamma \equiv \Theta_n - \Theta_s = \pi (R_\perp - r_i)(R_\perp + r_i) L \frac{H_{c1}^2}{8\pi} =$$
$$\delta b L \frac{H_{c1}^2}{8\pi} = \delta \frac{H_{c1}^2}{8\pi} A_b = \alpha A_b, \tag{5.45}$$

Here $\delta = R_\perp - r_i$, the radial width of the space between the cylinders, is the wall energy parameter introduced in Sec. 4.3; $b = 2\pi \left[(R_\perp + r_i)/2 \right]$ is the length of a notional S/N interphase boundary in the plane perpendicular to **H**; $A_b = bL$ is the area of this boundary; and $\alpha = \Gamma/A_b = \delta H_{c1}^2/8\pi$. The b-line, which is shown in Fig. 5.8 as a dash-dotted contour, can be viewed as an effective S/N interphase boundary.

We see that Γ is the excess energy caused by the presence of the S/N interphase boundary and therefore $\alpha(= \Gamma/A_b)$, the excess energy per unit area of this boundary, is the surface tension. Note that this is close to that how the S/N surface tension was introduced by H. London [138] and used by Landau in his models of the IS [154, 137]. On the other hand, the wall energy parameter $\delta = R_\perp - r_i$ is close to that proposed by Pippard: $\delta = \xi - \lambda_L$ [139].

Referring to Eq. (4.4), the total free energy of the specimen with one hole is

$$\widetilde{F}(T, H_0)_{1h} \equiv \widetilde{F}_{s0} - \int_0^{H_{c1}} M \, dH_0 = \widetilde{F}_n - \frac{H_c^2}{8\pi} V + (N_\perp - 1)\Theta_s + \Theta_n =$$
$$[\widetilde{F}_n - \frac{H_c^2}{8\pi} V + N_\perp \Theta_s] + \Gamma = \widetilde{F}(T, H_{c1}) + \Gamma, \tag{5.46}$$

where $\widetilde{F}(T, H_{c1})$ is the total free energy of the specimen without the hole at $H_0 = H_{c1}$, i.e., of the specimen in the MS, and $N_\perp$ is the number of micro-whirls (see Eq. (5.21)).

Thus, if $\aleph < 1$, or $\delta > 0$, or the surface tension α and, correspondingly, Γ is positive, the free energy of the cylindrical specimen without the hole

(N domain) is always (i.e., for any H_{c1}) less than that with the hole. Therefore, such a specimen stays in the MS all the way up to the thermodynamic critical field H_c; i.e., $H_{c1} = H_c$. At $H_0 = H_c$ the magnetic energy of this specimen E_m equals the condensation energy E_c and therefore superconductivity collapses at once all over the volume meaning that the specimen enters the N state via the phase transition of the first order. Therefore, materials with $\aleph < 1$ represent type-I superconductors, as it should be the case when the S/N surface tension is positive.

As we know (see Ch. 4), the field passes through a specimen of type-I superconductors only if its shape is non-cylindrical and the applied field $H_0 > H_c(1 - \eta)$. Herewith the specimen transitions to the IS in which (in case of sufficiently massive specimens) the field in the N domains $H = B = H_c$ and therefore the surface tension α takes the same form as that proposed by Landau (Eq. (4.8)).

The physical meaning of the S/N surface tension can be interpreted as follows. By passing the field through itself (i.e., creating an N-domain(s)), the superconductor gets rid of the induced currents in a certain volume (gain), and pays for this with condensation energy (losses) contained in a volume different from the first. In type-I materials the second volume is greater, meaning that the losses exceed the gain, so the superconductor does everything possible to reduce the losses, i.e., to reduce the area of the S/N interface when the appearance of N domains is unavoidable. At the same time, as we saw in Ch. 4, each specific specimen takes into account all other contributions in its free energy in the most rational way.

Now, what happens when $r_i > R_\perp$ or $\aleph > 1$? In this case, depicted in Fig. 5.8b, the same steps as above lead to Eq. (5.45), but now $\Gamma < 0$, or the surface tension α is negative. Then from Eq. (5.46) we see that the free energy of the specimen with the hole is *always* less than that without the hole. This means that the specimen should be in the MXS at any H_0 regardless how small this field is. Apparently, however, that the flux quantization prevents the appearing of the first hole until $H(= H_0$ in the cylindrical specimen) reaches a finite value H_{c1}. At this field the flux passing through the first hole equals Φ_0. In Ch. 4 we came to the same conclusion from purely thermodynamic reasoning.

To create the second hole, the applied field H_0 should be increased so that the total flux passing through the specimen equals $2\Phi_0$, and so on. At the same time, owing to the gain received due to the negative surface tension, the superconductor organizes the flux so to maximize the S/N interface area. The latter means that the flux passes through separate holes (the flux lines) with the flux quantum Φ_0 in each. The further increase of H_0 leads to appearance of the new flux lines, which geometry replicates the geometry of the micro-whirl lattice. Since the number of flux lines grows gradually, the transition to the N state is a continuous (second-order) phase transition occurring at $H_0 = H_{c2} > H_c$. As known, the outlined picture is consistent with the experiment (see, e.g., the right image in Fig. 4.1). Thus, materials with $\aleph > 1$ are

type-II superconductors as it should be the case when the S/N surface tension
is negative.

At H_{c2} the gain caused by the negative surface tension is exhausted. This
means that

$$\frac{H_{c2}^2}{8\pi}\pi r_i^2 = \frac{H_c^2}{8\pi}\pi R_\perp^2 \tag{5.47}$$

or

$$H_{c2} = \aleph H_c. \tag{5.48}$$

This formula is approximate because it does not take into account the
remaining condensation energy that supports superconductivity up to H_{c3}
(see Sec. 4.5). A similar formula was obtained by Abrikosov in the framework
of the GL theory [127].

Finally, what can be said about the geometry of filling the specimens with
the flux lines? The symmetry of the micro-whirls lattice suggests that the
first flux lines appear near the geometrical center of the specimen in the plane
transverse to **H**.

In majority of reported experimental observations the flux lines fill specimens starting from the edges, where concentration of defects is maximal.
However, in sufficiently pure specimens (single crystal $Bi_2Sr_2CaCu_2O_8$ [183]
and high purity Nb [184]) the observed picture looks similar to that expected
from the MWM^{37}.

5.2.8 Josephson effects

In 1962, 22-year-old Pippard student Brian Josephson published a seminal
theoretical paper on a tunneling effect in superconductors [123]. This paper
marked the opening of a vast field of research and applications, called weak
superconductivity or weakly coupled superconductors.

Analyzing what may happen in a junction between two superconductors
separated by a thin isolating layer, Josephson realized that at definite conditions a non-dissipating current ("supercurrent") can tunnel through the junction in the absence of voltage across it. The conditions are as follows. (a)
Cooper pairs in both superconductors (electrodes) should be in the ground
state, i.e. their wave functions $\Psi(= |\Psi|\exp(i\varphi))$ should be identical (i.e., equal
amplitudes $|\Psi|$ and the phases φ) throughout the volume of each separate elec-
trode. At the same time φ_1 (the phase of the wave function of Cooper pairs in
the first electrode) should differ from φ_2, while relationship between $|\Psi_1|$ and
$|\Psi_2|$ is not essential^{38}. (b) The coupling between electrodes should be weak.

---

^{37}Quoting Zeldov [183]: "... vortices seemed to appear mysteriously in the center of the
sample forming a vortex puddle with a dome shaped density surrounded by a vortex-free
region. With increasing field the dome grew from the sample center outward".

^{38}The invariance with respect to $|\Psi|$ means that the electrodes can be made of any (the
same or different) superconductors.

This means that the tunneling supercurrent should not disturb the wave functions of Cooper pairs in each of them.

The cause of supercurrent through such a junction (referred to as the Josephson junction) is the difference in the phases $\Delta\varphi = \varphi_1 - \varphi_2$[39].

Taking into account that the phase is a periodic function with the period 2π, Josephson predicted that at zero voltage across the barrier (junction) the supercurrent J_s is proportional to $\sin\Delta\varphi$, i.e.,

$$J_s = J_m \sin\Delta\varphi. \tag{5.49}$$

Here J_m, the maximum supercurrent, is referred to as the Josephson critical current. J_m is a characteristic of a given junction. An effect based on this equation is called the DC Josephson effect.

On the other hand, if a fixed voltage V is applied, Josephson predicted that it will continuously change the phase difference with the rate

$$\frac{\partial\Delta\varphi}{\partial t} = \frac{2eV}{\hbar}. \tag{5.50}$$

Accordingly, there will be an AC current through the junction with the linear frequency ν related to the voltage V as $2e/h = 483.6$ MHz/μV. This effects is referred to as the AC Josephson effect.

The applied field H_0 also changes the phase of the pairs while they pass the junction. As follows from Eqs. (A5.2) and (A5.3) in the footnote (39), this change $(\Delta\varphi)'$ is

$$(\Delta\varphi)' = -\frac{2e}{c\hbar}\int_0^l \mathbf{A}\cdot d\mathbf{l} = -\frac{2\pi}{\Phi_0}\int_0^l \mathbf{A}\cdot d\mathbf{l}, \tag{5.51}$$

---

[39] In quantum mechanics the electric current is a current of probability, which density is

$$\mathbf{j} = \frac{1}{2m}[\Psi\hat{\mathbf{p}}^*\Psi^* + \Psi^*\hat{\mathbf{p}}\Psi], \qquad (A5.1)$$

where Ψ is the wave function of the charge carriers having mass m, charge q and number density n; and $\hat{\mathbf{p}}(= -i\hbar\nabla - q\mathbf{A}/c)$ is the operator of the kinetic linear momentum these carriers [32, 49].

After substituting $\Psi = |\Psi|\exp(i\varphi) = \sqrt{n_{cp}}\exp(i\varphi)$, the expression for the density of supercurrent (the current transported by Cooper pairs) takes the form

$$\mathbf{j}_s = \frac{\hbar n_{cp}(2e)}{(2m)}(\nabla\varphi - \frac{2e}{c\hbar}\mathbf{A}). \qquad (A5.2)$$

Comparing $\mathbf{j}_s$ with the definition of the current density $\mathbf{j} = qn\mathbf{v}$, the expression for the velocity of Cooper pairs is

$$2m\mathbf{v} = \hbar\nabla\varphi - 2e\mathbf{A}/c. \qquad (A5.3)$$

Here the first term $(\hbar\nabla\varphi/2m)$ is the translational (drift) velocity of the pairs directed along the phase gradient; and the second term $(-2e\mathbf{A}/2mc)$ is the linear velocity of the field-induced circular motion of the pairs, causing the diamagnetic response (see Eqs. 2.15–19 and 5.11).

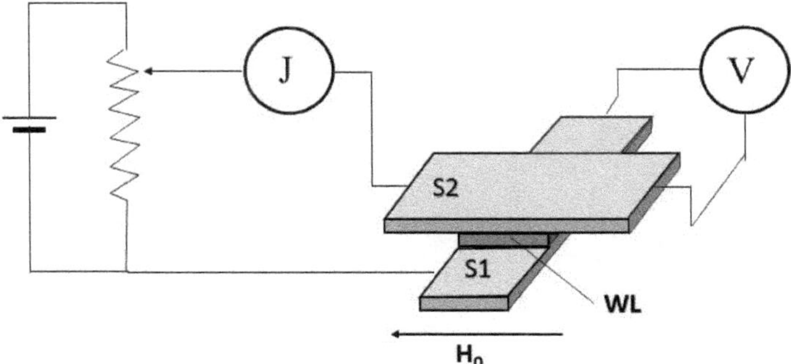

Figure 5.9 Schematic diagram of measuring the current-voltage characteristic of a Josephson junction. $S1$ and $S2$ are superconducting electrodes; WL (weak link) is a thin ($\sim$ 3 nm) isolating barrier, originally proposed by Josephson; it can also be non-superconducting metallic film less than 10 nm thick. H_0 is the applied field.

where l is the length of the circular path (arc) that the orbiting paired electrons travel while passing through the junction, dl is the vector element of this path, $\mathbf{A}$ is the vector potential of the field $\mathbf{H_0}$ (equal to $\mathbf{B_0}$ inside the junction), and Φ_0 is the flux quantum.

Respectively, the total phase difference in the field presence is

$$\Delta\Phi = \Delta\varphi + (\Delta\varphi)'. \tag{5.52}$$

Josephson's equations have been thoroughly verified and confirmed experimentally. Although the original predictions concerned tunnel junctions, it turned out that Josephson's equations are equally applicable to a wide class of junctions of various designs, in which the junction material could be not only an insulator, but also a normal metal, or the junction can be just a narrow neck of the same superconducting material as the electrodes [185, 85]. A schematic diagram for measuring the DC Josephson effect and a typical voltage-current characteristic for a tunneling junction are shown in Figs. 5.9 and 5.10.

Josephson junctions are used in extremely sensitive voltmeters and magnetometers and in the most accurate measurements of the ratio h/e. In a few years after Josephson's discovery the voltage standard based on his effect was developed; currently this is a transportable devise generating a programmable stable voltage with an uncertainty $\leq 1\cdot 10^{-10}$ [187].

Theoretical, experimental and applied aspects of Josephson effects are discussed in a number of monographs (e.g., [185, 188, 189]), which contain references to original works. In particular, it was found that the supercurrent in the junctions is transported by so-called Josephson vortices, superconducting units enclosing the flux quantum Φ_0, which progress slowly through the medium of the junction in the direction of the phase gradient $\nabla\varphi$ [85, 191, 190].

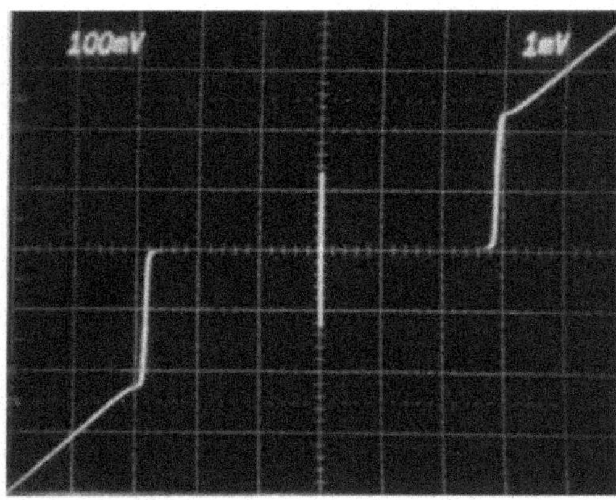

Figure 5.10 I-V characteristic for a tunneling Nb/AlOx/Nb junction at 4.2K (vertical axis: 1mA per division; horizontal axis: 1mV per division). The zero voltage current is clearly seen. After Shoji [186]; reprinted with permission from Elsevier.

Originally, Josephson derived his equations from quantum-mechanical calculations using Bogolubov's formalism and time-reversal symmetry to justify the sinphaseness of the wave functions [123]. Soon after, he offered a more transparent derivation considering the complex order parameter of the GL theory with particular emphasis on its imaginary part (the phase) [192]. Alternative derivations based upon general principles and the flux quantization were proposed by Feynman [49] and Bloch [193].

In the MWM, the DC Josephson effect stems from zero entropy of the S phase, which means that the induced currents in all Cooper pairs of a given specimen (electrode) are in phase. As we have seen, this follows from the fact, that the pairs are in the ground state all over the temperature and field ranges of the S state (see Secs. 5.2.1-5.2.3).

Therefore, as soon as the link between the electrodes is weak (i.e., Josephson's condition (b) is met), the zero-voltage supercurrent J_s will flow, provided the phase difference $\Delta\varphi$ in the induced currents of the linked electrodes is not zero.

Since the superconducting current is a collective motion of the micro-whirls, the current in the junction represents a slow (the current is small, ~ 1 mA) diffusion-like motion of the whirls through the junction driven by the phase gradient $\nabla\varphi = \Delta\varphi/d$, where d is the width of the junction. Mathematically, this process is described by Eq. (A5.3) in footnote (^{39}) for the pairs' velocity at zero magnetic field (i.e., with $\mathbf{A} = 0$), or by the Josephson equation Eq. (5.49) for the total current.

The DC electric field $\mathbf{E} = -\nabla V$ produced by the voltage V across the junction accelerates the whirls while they pass the junction. Respectively, if at $V = 0$ Cooper pairs on both sides of the junction had the same energy (the pairs' assembles on both sides were at zero temperature), now each pair in the whirl arriving to the other side has an extra energy

$$E_{ex} = 2eV. \tag{5.53}$$

To thermolize ("to became as all"), the arriving pairs have to get rid of this energy. However, due to zero temperature of the pairs there, they cannot communicate with them. Hence, the only way to thermolize, is to radiate this energy, that is to emit photons with the frequency ν equal to

$$\nu = \frac{E_{ex}}{h} = \frac{2eV}{h}. \tag{5.54}$$

But electromagnetic radiation means the presence of AC current of the same frequency [1]. Hence, at $V \neq 0$ the supercurrent through the junction is

$$J_s = J_m \sin\left(\frac{2eV}{\hbar}t + \Delta\varphi_0\right) = J_m \sin\left(2\pi\frac{Vc}{\Phi_0}t + \Delta\varphi_0\right), \tag{5.55}$$

where $\Delta\varphi_0$ is the phase difference at $V = 0$.

This is equivalent to Eq. (5.50).

Important to stress that the radiation is coherent, which was confirmed for the first time by Shapiro [194][40]. The phase-coherence of the AC Josephson currents lays in foundation of the SQUIDs (superconducting quantum interference devices), used in the precision magnetometry.

Now, let us look a bit closer what happens when a magnetic field H_0 is applied. For simplicity, consider the case when $\mathbf{H}_0$ is parallel to the junction sides, e.g., as shown in Fig. 5.9.

Eq. (5.51) is often referred to as the gauge-invariant phase difference. However, as we know (see Sec. 1.2.8) the gauge invariance is inapplicable to superconductors[41] and the only appropriate gauge of the vector potential is the circular one (Eq. (1.43)).

Therefore, since in our case $\mathbf{H}_0 (= \mathbf{B}_0)$ equals the field intensity $\mathbf{H}$ inside the electrodes, the vector potential has the form

$$\mathbf{A} = \frac{\mathbf{H} \times \mathbf{r}_i}{2}, \tag{5.56}$$

where r_i is the radius of the field induced currents forming the micro-whirls.

Now, taking into account that the vector $(\mathbf{H} \times \mathbf{r}_i)$ is parallel to $d\mathbf{l}$, and the fact that each whirl carries the flux quantum Φ_0 (which means that $H\pi r_i^2 = \Phi_0$), Eq. (5.51) is rewritten as

$$-\frac{2\pi}{\Phi_0}\int_0^l \frac{\mathbf{H} \times \mathbf{r}_i}{2}d\mathbf{l} = -\frac{H\pi r_i^2}{\Phi_0}\int_0^l \frac{dl}{r_i} = (\Delta\varphi)', \tag{5.57}$$

---

[40]In the $V - J$ characteristics of Josephson's junction (Al/Al$_2$O$_3$/Sn) subjected to microwave irradiation Shapiro observed a step-like changes in the current at $V = \nu_r(h/2e)n = (\Phi_0/\lambda_r)n$, where ν_r and $\lambda_r(= c/\nu_r)$ are the frequency and wavelength of the microwave irradiation, and $n(= 0, \pm1, \pm2, ...)$ is an integer.

[41]Indeed, $\mathbf{A}$ of different gauges for the same field H_0 would result in different $(\Delta\varphi)'$ and therefore in different current, which is unphysical.

where $(\Delta\varphi)'$ is the additional phase change acquired by paired electrons, while they pass the junction. Thus the MWM discloses the physical meaning of otherwise "mysterious" [49] phase of the wave function. Herewith, the physically significant (measurable) quantity is not the phase φ itself, but the phase difference $\Delta\varphi$ of the wave functions (that is, of the induced currents in Cooper pairs) in the weakly coupled superconductors, as predicted by Josephson.

Finally, we note that the Josephson vortices represent the micro-whirls of the MWM, the cylindrical structures of densely packed circular currents induced in Cooper pairs, with the superconducting flux quantum Φ_0 enclosed in each of the whirls. Inside the latter, the $\mathbf{H}$ field is compensated by the field created by the current loops, i.e. $\mathbf{B} = 0$. Accordingly, when the micro-whirls are probed using a scanning tunneling microscope (STM), the tip of the microscope "sees" the circular current on the whirl's edge and no current in its core since the currents due to orbital motion of the paired electrons are compensated (see Eq. (5.7)). This can make an impression that the whirl core is real (normal). STM images of such kind for Josephson vortices were indeed reported by Rodichev with coauthors [190].

5.2.9   Flux lines *vs* Abrikosov vortices

As known (see Sec. 4.4), the magnetic moment of type-II superconductors in the MXS being negative increases (decreasing in magnitude) with increasing applied field due to the flux lines entering the specimen. Let us consider the structure of these lines.

A diagram illustrating the standard interpretation of the flux lines, often called Abrikosov vortices, is shown in Fig. 5.11. According to this interpretation, induced currents in each vortex run around its core, the size of which is significantly smaller than the size (radius) of the currents. The later are directed counterclockwise, when viewed from the field tip, i.e. the vortices are paramagnetic. Such an interpretation cannot but raise questions, because currents induced by magnetic field in a singly connected specimens (regardless superconducting or not) are *always* diamagnetic. On the other hand, numerous experiments confirm the vortex structure of the MXS (see, e.g., the right image in Fig. 4.1)

The flux line in the MWM is schematically shown in Fig. 5.12. The minimum magnetic flux (the flux quantum Φ_0) passes through the specimen when superconductivity ceases in the volume of one micro-whirl. Thus, there will be a non-superconducting hole (open circle) in the network of the superconducting micro-whirls (shown in gray).

Current around the hole is an effective paramagnetic current (shown by the dashed line) formed by the induced diamagnetic currents in the micro-whirls. Correspondingly, contribution of each such a hole into the specimen magnetic moment is positive in agreement with experiment.

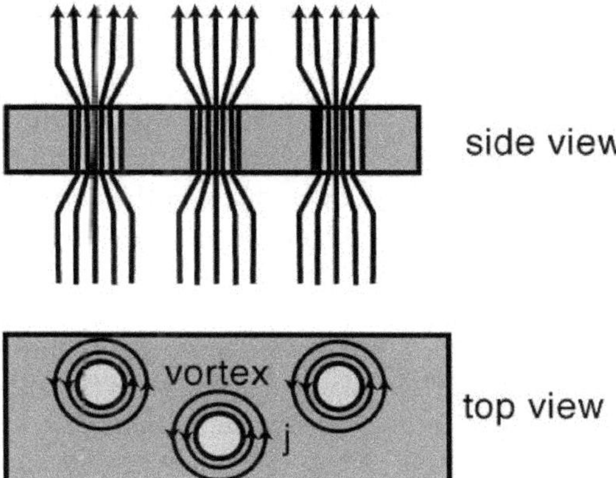

Figure 5.11   A diagram of Abrikosov vortices copied from Google's images of vortices in superconductors. The field-induced not-dissipating current with the volume density **j** flows counterclockwise around the vortex core, i.e., the current is paramagnetic. The charge carriers are Cooper pairs not experiencing resistance. Note that the Lorentz force driving the pairs is directed outward from the center, and therefore such a current is possible only if either the charge of Cooper pairs is positive or their mass is negative. Alternatively, such a current increases the field passing through the vortex, and therefore increases the specimen free energy, which is contrary to thermodynamics.

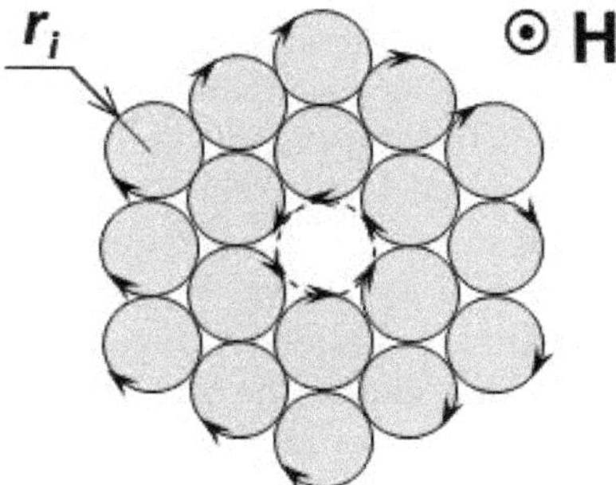

Figure 5.12   Flux line in the MW model. Solid circles with arrows represent the current J_{cp} in the cross section of the micro-whirls shown in gray; r_i is the rms radius of these currents. A central open circle, the core of the flux line, is a hole in the network of the diamagnetic micro-whirls. An effective paramagnetic current surrounding the core is shown by the dashed line. It is formed by the diamagnetic currents induced in the pairs surrounding the core. After Kozhevnikov [116], reprinted with permission from Springer Nature.

One can assign and calculate the energy per unit length (length tension) of the flux lines. It has the same form as that of Abrikosov vortices at H_{c1} (see Problem 5.3).

One of the main differences between the standard and the micro-whirls' interpretations of the flux lines is that in the latter the area of passing flux is primarily restricted by the rms radius r_i, while in the former it is much greater (the current, called the screening current, decays exponentially [85]). Due to this, the currents in Abrikosov vortices overlap with the fields passing through their neighbors, which results in the interaction between these vortices. This conflicts with thermodynamics, since the interaction increases free energy. Recall that the free energy of the specimens in the MXS (and other equilibrium states) does not contain terms responsible for interaction.

On the contrary, in the MWM the flux lines do not interact (due to the absence of interaction between the micro-whirls), in agreement with thermodynamics and experiments (see Sec. 4.4).

5.2.10 Thermoelectric effects

As discussed in Sec. 3.1.2, experiments demonstrate the absence of thermoelectric effects in superconductors. This fact prompted Gorter and Casimir to introduce the zero entropy postulate in their two-fluid model. The origin and physical meaning of this postulate is explained in the MWM: owing to the perfectly organized micro-whirl structure of the field-induced currents in Cooper pairs, the entropy of the S phase is zero at all temperatures and fields of the S state. This automatically entails the absence of the entropy transport which generates thermo-e.m.f. caused by a temperature gradient over the specimen. However, in any superconductor majority of conduction electrons are unpaired (normal) and they must respond to the temperature gradient. Therefore, the absence of thermo-e.m.f. in superconductors requires an addition explanation. A qualitative scenario could be as follows.

In a normal metallic wire the Seebeck voltage (thermo-e.m.f. $\mathcal{E}$) arises due to a heat flow J_Q from the hot end of the wire (with temperature T_1) to the cold one (with temperature T_2) at the absence of electric current. In terms of entropy, this effect represents an entropy current J_S from the cold end to the hot one with a simultaneous production of entropy within the wire[42]. The entropy current leads to the appearance of a potential difference between the ends of the wire and, accordingly, of an electric field, which stops the current and brings the system into equilibrium[43] [195].

---

[42]This description reflects the fact that the entropy entering into the wire per unit time from the hot reservoir (equal to J_Q/T_1) is less than the entropy exiting the wire to the cold reservoir (equal to J_Q/T_2) because $T_1 > T_2$. The entropy balance is maintained by the entropy production within the wire.

[43]Recall that the Seebeck effect, as well as the Peltier and Thomson effects, are phenomena that occur under conditions of thermodynamic equilibrium.

The Seebeck effect is interpreted as a result of diffusion of electrons from the cold end to the hot one caused by the difference in densities (see, e.g., [196]).

In superconductors the situation is different. Due to zero entropy and, accordingly, zero T_{cp}, the ensemble of Cooper pairs is insensitive to both the specimen temperature T and its gradient ∇T. However, as we well aware of (see Secs. 3.2.1 and 5.2.3), the pairs' density depends on T. Hence, the temperature gradient in the specimen creates a gradient in the density of the pairs. Specifically, the pairs' density in the warm end is less than that in the cold end. Respectively, the density of normal electrons is higher in the warm end, i.e., there is a gradient in the density of the normal electrons directed from the warm to the cold end. Hence, the gradient in the density of the normal electrons caused by the presence of Cooper pairs is directed *opposite* to that caused by the different temperatures. The experimental fact $\mathcal{E} = 0$ indicates that these gradients cancel each other. This results in the absence of entropy current and therefore in the absence of the entropy production in the ensemble of normal electrons in superconductors.

More detailed discussion of this scenario requires a quantitative theory of the thermoelectric field in the normal metals, which is missing [41, 197].

As was first shown by W. Thomson (Lord Kelvin) [195], the equilibrium thermoelectric effects are interrelated and the absence of the Seebeck effect leads to vanishing the Peltier and Thomson effects[44]. This is consistent with experimentally established fact of the absence of all equilibrium thermoelectric effects in superconductors [15].

5.2.11   Magnetization by rotation (London moment)

Shortly before the Meissner effect was discovered, Becker, Heller and Sauter proposed a so-called acceleration theory of magnetic properties of superconductors, assuming that the latter are perfect conductors in which electricity is transported by totally free electrons [198][45].

For a stationary FC specimen (the authors considered a solid sphere) subjected to a magnetic field $\mathbf{H}_0 < \mathbf{H}_c$, Becker et al. found that the field in the specimen interior would be frozen. This means that as $\mathbf{H}_0$ changes, the field inside remains the same as it was when the material became superconducting due to the induced resistanceless circumferential surface current acting as a screen. The screening arises due to complete or partial compensation of the applied field by the field $\mathbf{H}_i$ produced by the surface current inside the specimen. In the ZFC case, compensation is complete, implying that $\mathbf{H}_i$ is equals

---

[44]The Peltier effect, the release or absorption of heat when current is passed through a junction of two different metals, represents a reversed Seebeck effect, so the absence of the latter automatically entails the absence of the former. The Thomson effect is a continuous version of the Peltier effect. Hence, it should also disappear in superconductors.

[45]A description of this theory in English is available in [15, 199].

to $-\mathbf{H}_0$. Thus, the authors repeated Maxwell's solution of the same problem: "If the sheet [of infinite conductivity] forms a closed or an infinite surface, no magnetic actions which may take place on one side of the sheet will produce any magnetic effect on the other side" [1].

For a rotating solid sphere cooled in a zero field before rotation begins, Becker et al. predicted the appearance of a surface current caused by a small number of "superelectrons" which lag behind near the surface from their counterparts in the bulk. The latter, in turn, move in phase with the ionic lattice. Herewith a distribution of the surface current in the rotating specimen is the same as that of the current induced in the stationary case by the applied field, but is directed in opposite direction. The rotation-induced surface current generates the field $\mathbf{H}_i$ in the interior equal to

$$\mathbf{H}_i = -\frac{2mc}{e}\boldsymbol{\omega}_r, \qquad (5.58)$$

where $\boldsymbol{\omega}_r$ is the angular velocity of rotation.

On the other hand, no surface current arises at cooling an already rotating specimen. Accordingly, no $\mathbf{H}_i$ arises in such case.

Soon thereafter, due to apparent contradiction of the acceleration theory with the Meissner effect, the London brothers proposed a modified version of this theory. To account for the Meissner effect, they postulated the independence of the screening on the initial conditions (see Sec. 3.2.2). Considering the rotating specimen (also a ball) F. London took over the result of Becker et al., but extended it to the rotating specimens (i.e., to the specimens that are cooled in a rotating state), referring to the Meissner effect [35].

The field $\mathbf{H}_i$ in Eq. (5.58) is often called the London field, and the effect of magnetizing a superconductor by rotation, as the London moment. Apart from the original acceleration theory [198], a number of alternative derivations of $\mathbf{H}_i$, have been proposed (e.g., [200, 201]). As a rule (which includes F. London's calculations [35])) Eq. (5.58) is an approximation valid for the massive singly connected spheroidal specimens, i.e., in the limit of negligible thickness of the surface layer containing the screening current. This surface layer forms a so-called excluded volume, which is discussed in the next section.

The rotation-induced field is very small: at $\omega_r \sim 10^2$ Hz this field is of the order of 10^{-5} G. Therefore, in experimental tests of this effect, the terrestrial field is thoroughly compensated (e.g., in the experiment of Tate et al. [202], the Earth's field was compensated down to 5×10^{-8} G). There are also strict requirements for the sample geometry and purity[46].

The first successful experiment on the London moment was reported by Hildebrandt [64]. The samples were two lead cylindrical shells with an aspect

---

[46]As a rule, the authors attach great importance to the purity of the samples. However, in [203] it was shown that the abundance of defects in ceramic (disc-shaped) high-T_c superconductors does not play an essential role in measurements of the London moment.

ratio (diameter to length) of $2/3$[47]. Results obtained were in agreement with Eq. (5.58) within the experimental error (7%).

Subsequently, the experiments were performed with type-I and type-II superconductors, including high-T_c materials[48]. The studied samples were solid and hollow cylinders of different dimensions, rings and thin-film spherical shells. In all cases authors reported an excellent (within the error bar) agreement with the London theory[49]. Basing on this fact, Tate with co-authors used Eq. (5.58) for very delicate measurements of the mass of Cooper pairs. The effect of London moment underlies the design of a high-precision cryogenic gyroscope used in a space mission devoted to verification of a fine relativistic effect [204].

Overall, the correctness of Eq. (5.58) is beyond doubt at least for the sample geometries tested in experiments. The question is why this is so, since, as we know, neither the acceleration theory of Becker, Heller and Sauter, no its modification by the Londons are adequate.

In the MWM the magnetization by rotation of singly connected superconductors represents a reversed effect of rotation by magnetization in superconductors of I. Kikoin and Goobar or the Barnett effect in diamagnets anticipated by I. Kikoin [62] (see Sec. 2.2 and problems 5.4 through 5.6). In its turn, the field $\mathbf{H}_i$ in Eq. (5.58) is the Barnett field $\mathbf{H}_B$ discussed in Sec. 2.1. This field inside a spinning spheroidal specimen is

$$\mathbf{H}_B = -\frac{2mc}{e}\boldsymbol{\omega}_r. \tag{5.59}$$

Important to emphasize that, unlike the London field, the Barnett field is uniform throughout the entire specimen without any excluded volume. Also, the Barnett field is parallel to $\mathbf{H}_0$ applied to the motionless specimen. Recall that one of the most important differences between the acceleration theories and the MWM is the absence of a circumferential (continuous or total) current in the latter, which provides the independence of the magnetic properties on the initial conditions and compliance of this model with the laws of physics.

Like in rotating ferromagnetics in the Barnett effect, the Barnett field causes magnetization in rotating superconductors. Therefore, as it was pointed out by Barnett [55], $\mathbf{H}_B$ is the *intensity* of the rotation-generated magnetic field by definition (Eq. (1.15)).

In the case of specimens of the cylindrical geometry, due to zero demagnetizing factor η, $\mathbf{H}_B = \mathbf{H}_{ext}$, where $\mathbf{H}_{ext}$ is the external field produced by the magnetized specimen. Since $\mathbf{H}_i$ in the acceleration and London theories

---

[47]The author noted the inconsistency of the shape of his sample to that required by the London theory.

[48]The results of measurements of the London moment in high-T_c superconductors of Verheijen et al. [203] were the first evidence that superconductivity in these materials does not differ from that in "conventional" low-T_c superconductors.

[49]In most experiments the sample magnetic moment M vs ω_r was measured. The principal quantity to compare with the theory was the slope $dM/d\omega_r$.

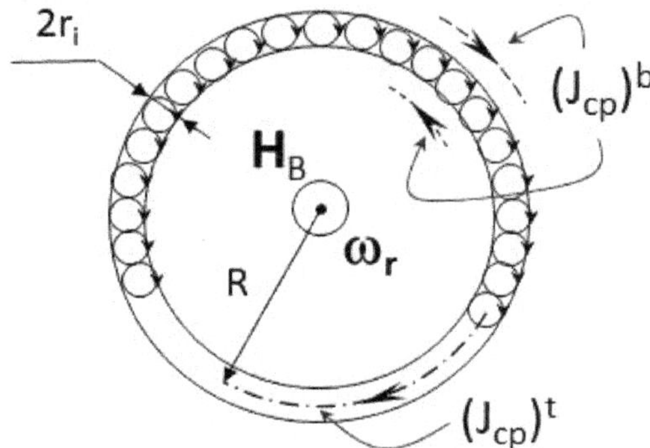

Figure 5.13 Cross section of a thin-wall superconducting tube magnetized by rotation (MWM model). Velocity of rotation $\boldsymbol{\omega}_r$ and the rotation-induced Barnett field $\mathbf{H}_B$ are parallel and directed toward the reader. Small circles are cross sections of the micro-whirls with rms radius r_i, small arrows there indicate direction of the current in Cooper pairs J_{cp} induced by the Barnett field. Dashed arcs with arrows designate direction of the bound currents J_{cp}^b on both sides of the ring. Dot-dashed arc with arrow indicates direction of the total current J_{cp}^t caused by the changing magnetic flux inside the tube aroused due to the Barnett field $\mathbf{H}_B$. The tube radius $R \gg r_i$ and the tube length (perpendicular to the page) $L \gg r_i$. Relation between R, L and the wall width is not important (see text).

(Eq. (5.58)) coincides with the Barnett field $\mathbf{H}_B$ (Eq. (5.59)), the magnetic moments calculated in the MWM and in these theories coincide also, provided that the excluded volume is negligible (see problems 3.2-3.). This explains the consistency of the acceleration and the London theories with experimental results on the London moment obtained with solid cylinders.

Less trivial is the case of magnetization by rotation of multiply connected samples, such as hollow cylinders, rings and spherical shells. Recall that the principal difference between the singly and multiply connected superconductors is the possibility of the appearance of a total current in the latter (see Sec. 3.1.4).

Consider a long ZFC hollow cylindrical sample rotating about its longitudinal axis. Cross section of such a sample is schematically shown in Fig. 5.13. The Barnett field $\mathbf{H}_B$ is parallel to the generating line of this cylinder, the Cooper pairs in the wall precess about the lines of this field and form the ordered micro-whirl's structure as shown in the figure (the number of rows of micro-whirls is determined by the wall thickness; in the figure this thickness is equal to $2r_i$). Since $\mathbf{H}_B$ is the intensity of the magnetic field inside the wall, the field *outside* the wall has the same intensity $\mathbf{H}_B$ due the boundary condition for H_t (Eq. 1.18). Our cylinder is in free space, therefore the induction $\mathbf{B} = \mathbf{H}_B$ on the both sides from the wall (inside the wall $\mathbf{B} = 0$). This means that as soon as the cylinder starts spinning, the flux of the magnetic field in its opening changes, thus causing the induction e.m.f. in accord with the Faraday law. In turn, the e.m.f. causes a total current designated in the figure as $(J_{cp})^t$.

Now, since the micro-whirls are ordered (the induced currents inside the wall are structured in the same way as in the film which magnetization curves are shown in Fig. 3.6), the total current flows without resistance and therefore the flux in the opening is locked. This means that the induction within the opening of the ZFC hollow cylinder stays zero, as this would be if the cylinder were solid (singly connected). Respectively, the magnetic moment of our hollow cylinder is the same as that of the solid cylinder with the field intensity $\mathbf{H}_B$ inside it.

The reader may object, that the zero-field cooling implies a presence of although small but never zero residual terrestrial field. Therefore the locked flux within the shell is not zero, while it is exactly zero inside the solid cylinder. This is true, but the induced total current always compensates the Barnett field $\mathbf{H}_B$ within the opening regardless on the initial condition. Respectively, the magnetometer will record the same magnetic moment for the ZFC and FC samples, which is fully consistent with experiments, in particular, with results of still the most comprehensive study of this effect by Brickman [205]. As an example, experimental data on the London moment reported in this work are shown in Fig. 5.14.

Another possible objection can be related to the fact that we consider a long cylinder, while in experiments this is not always so. In such cases the magnetometer is usually calibrated with respect a solid sample of the same dimensions (see, e.g., [203]). This makes the cylinder length insignificant since the induced total current compensates the bound current on the inner side of the ring or shell. Since the bound current is induced by the Barnett field, the field extracted from these measurements is automatically equal to the Barnett field. For the same reason, the wall width is insignificant also, provided it exceeds the minimum possible thickness (see Sec. 5.2.13). At the same time, similar as it takes place in the stationary ring (Sec. 3.1.4), the sample magnetic moment is equal to the *sum* of the moments due to the induced total current and the bound currents induced in Cooper pairs.

The measurements of the London moment are equivalent to the measurements of the gyromagnetic ratio γ in the Barnett effect. The g-factor of the total circular current induced in a hollow sample is unity by definition. Therefore, the experimentally observed agreement of the rotation-induced field with the Barnett field means that the g-factor of the microscopic currents in Cooper pairs is unity as well. This is consistent with experiment of I. Kikoin and Goobar (see Ch. 2) and serves as an additional confirmation of the fact that spins of the paired electrons are mutually compensated in *all* superconductors.

Now, let us see what is the field and corresponding magnetic moment in non-cylindrical bodies, for example in the ball, considered by Becker, Heller and Sauter and by F. London. Recall that according to F. London, the field inside a spinning singly connected spheroidal specimen of any geometry

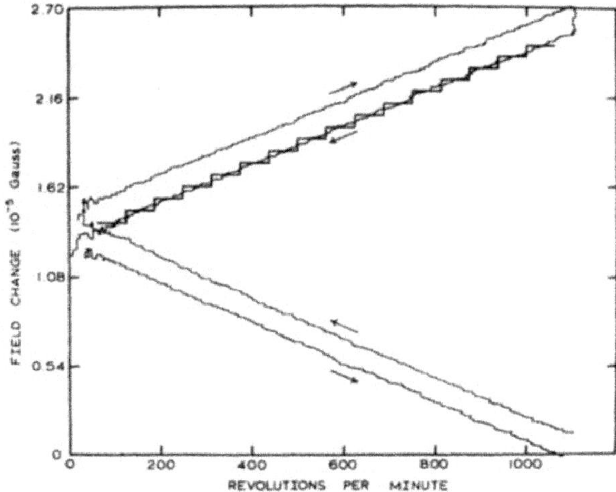

Figure 5.14 Example of London-moment data reported in [205]. The sample is a single-crystal tin rod 60 mm long and 6 mm in diameter; temperature 1.20 K. The arrows indicate direction of speed change. The individual plots are spaced by changing the y-axis zero of reference. The slop in the graphs matches $2mc/e$ with an error 3%. The superimposed steps are calculations on a theory in which the London moment is produced by Abrikosov vortices; clearly the experiment does not support that theory. Reprinted with permission from APS.

equals $-\mathbf{H}_0$[50], where $\mathbf{H}_0$ is the field applied in the stationary case. Evidently, this is not the case in the MWM: the Barnett field in the spinning specimen with demagnetizing factor η is equal to $\mathbf{H}_0/(1-\eta)$; for the ball this means that $\mathbf{H}_B = 3\mathbf{H}_0/2$ (see Sec. 1.2.5).

Note an interesting situation taking place in a superconducting thin-walled spherical shell (as well as in other spheroidal shells; a long hollow cylinder is a particular case of these shells). One can show (see problems 3.4-6.), that the magnetic moment of such a shell is the same as the moment of the solid sphere of the same size. Referring to Eq. (1.30) the moment of the rotating sphere is equal to

$$\mathbf{M} = -\frac{V}{4\pi}\mathbf{H} = -\frac{V}{4\pi}\mathbf{H}_B = -\frac{R^3}{3}\mathbf{H}_B = -\frac{R^3}{2}\mathbf{H}_0, \qquad (5.60)$$

where R is the sphere radius and $\mathbf{H}_0$ is the applied field in the stationary case, which produces the same moment as the Barnett field in the rotating ball.

Now, if we replace the rotating sphere by the rotating spherical shell of the same radius and with the same magnetic moment (i.e., the shell and sphere rotate with the same ω_r), the field inside this shell is $\mathbf{H}(=\mathbf{B}) = -\mathbf{H}_0$, which

---

[50]In the acceleration theory this relationship is valid only for ZFC specimens.

is 1.5 times less in magnitude and directs opposite to the field intensity inside the solid sphere. As one can see, this field is the same as in the London theory! This seemingly strange result is just a confirmation that in the London theory a solid specimen is equivalent to it shell (see Sec. 3.2.2). The field inside a shell can be measured by μSR using a film sample deposited on, e.g., a glass ball. The **H**−field in a solid sphere can be measured by NMR and ESR (see Sec. 5.2.15).

Concluding this section, we note that the existence (the possibility of observing) of the direct and inverse effects of I. Kikoin and Goobar suggests that the magnetic resonance caused by the precession of Cooper pairs (i.e., an electron resonance with $g = 1$) may exist also.

5.2.12 Penetration depth, excluded volume and a thermodynamic argument of H. London

In 1947, Heinz London put forward a thermodynamic argument [206][51] showing a controversy of the notion of penetration depth, the central concept of the London and other standard theories. Let us examine this argument.

Consider a long superconducting (for definiteness type-I) rod of radius $R \gg \lambda_L$ in a parallel field $\mathbf{H}_0$, i.e., a cylindrical specimen in the MS shown in Fig. 3.8. The Maxwell relation for this case reads[52] (see, e.g., [16])

$$\left(\frac{\partial S}{\partial H_0}\right)_T = \left(\frac{\partial M}{\partial T}\right)_{H_0},\tag{5.61}$$

where S and M are the entropy and magnetic moment of the specimen, respectively, and T is its temperature.

As mentioned in Sec. 3.2.2 (see footnote (30) on p. 75), in the London theory the magnetic moment of this specimen is

$$M = -\frac{H_0}{4\pi}(V - V_p) = -\frac{H_0}{4\pi}(V - \lambda_L A_s),\tag{5.62}$$

where V_p and A_s are the volume and surface area of the layer with the effective penetration field, respectively, and λ_L is the London penetration depth. Recall that according to the London theory in a specimen with a plane boundary (in particular, in the cylinder under consideration) λ_L is the constant of an exponential decay of the induction $B(z)$, where z is the distance from the surface; the penetration depth λ in this theory is infinite and its effective value $\lambda_{eff} = \lambda_L$. In turn, λ_{eff} is defined as a width of the surface layer containing the flux of the induction of the penetrating field assuming that

---

[51]This argument is considered in [128] and [35].

[52]In general case, i.e., for any ellipsoidal body in an arbitrary oriented field this relationship is $(\nabla_{\mathbf{H}} S)_T = (\partial \mathbf{M}/\partial T)_H$.

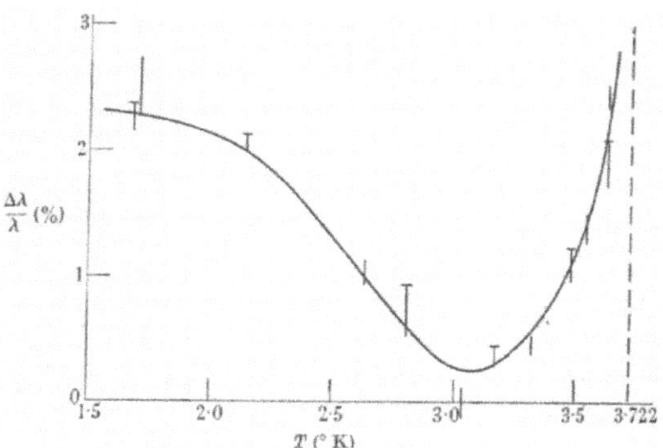

Figure 5.15 Smoothed experimental data on the change in the London penetration depth between zero and critical field, as a function of temperature (Pippard, 1950 [128]). Vertical lines designate estimates of experimental error. The measurements were performed with a microwave resonator technique at frequency 10 GHz. Reprinted with permission from the Royal Society London.

B is constant and equal to $B_0(= H_0)$, i.e., $B_0\lambda_{eff}A_s/L = (A_s/L)\int B(z)dz$, where L is the length of our cylindrical specimen and A_s/L is its perimeter.

The volume V_p represents the excluded volume possessing a quite peculiar property: it does not contribute into the specimen magnetic moment, but the current in this volume creates this moment.

Next, neglecting thermal expansion and taking into account that λ_L depends on T (in the London theory λ_L is a function of temperature only), from Eq. (5.62) it follows

$$\left(\frac{\partial S}{\partial H_0}\right)_T = \frac{A_s H_0}{4\pi}\left(\frac{\partial \lambda_L}{\partial T}\right)_{H_0}. \tag{5.63}$$

Therefore, since $(\partial\lambda_L/\partial T)_{H_0} \neq 0$, the specimen entropy S depends on the field at constant temperature. Important to point out that this entropy is contained entirely within the penetration layer and its density (calculated as $(S - S_0)/V_p$, where S_0 is the entropy at zero field) is *not small* [35, 15, 128].

According to the two-fluid model, the entropy of a superconductor in the MS is caused by the change of n_s[53]. Hence, the field dependence of S implies that n_s depends on the field at constant temperature. Therefore, $\lambda_L(\sim 1/\sqrt{n_s})$ should *depend* on the field. This is the essence of H. London's argument.

This argument motivated Pippard to verify the field dependence of the penetration depth using his newly developed microwave technique [128]. The reported results are reproduced in Fig. 5.15. Pippard noted that the percentage values presented in the graph are probably the maximum values, but their

---

[53] Actually, the following is not quite a lawful extension of the two-fluid model, where n_s changes only with temperature.

verification requires a technique allowing to perform similar measurements at a low frequency[54]. Pippard concluded that the penetration depth can be considered as independent of the field. As mentioned above, a quarter of a century later Pippard noted that this problem was still unsolved [129].

Interesting, that analyzing the results on the field dependence of λ_L[55], Pippard concluded that F. London's hypothesis of the long-range ordering of the S phase is consistent with his data. From here Pippard came to the idea of nonlocality of superconducting charge carriers, which in turn paved the way for the BCS theory.

But let us return to the penetration depth with the excluded volume and see what can be said about the field dependence of λ_L basing solely on thermodynamics.

By definitions of the condensation energy $E_c(T)$ and the thermodynamic critical field $H_c(T)$ (Eqs. (4.2,3)), the difference between the total free energies of the specimen in the S ($\widetilde{F}_s$) and the N ($\widetilde{F}_n$) states in zero field is

$$\widetilde{F}_{n0} - \widetilde{F}_{s0} = \widetilde{F}_n - \widetilde{F}_{s0} = E_c(T) = V\frac{H_c^2(T)}{8\pi}, \qquad (5.64)$$

where subscript 0 designates zero field.

Next, by definition of the total free energy (see Ch. 4), $\widetilde{F}_s$ of a cylindrical specimen in the field $\mathbf{H}_0$ at constant T is

$$\widetilde{F}_s(T, H_0) = \widetilde{F}_{s0} - \int_0^{H_0} \mathbf{M}d\mathbf{H}_0 = \widetilde{F}_n - V\frac{H_c^2}{8\pi} + V\frac{H_0^2}{8\pi}. \qquad (5.65)$$

Here we also used the definition of the MS according to which $\mathbf{M}$ in this state does not depend on temperature at constant field, as it takes place in all diamagnetics (see, e.g., Fig. 3.6).

From Eq. (5.65), as we saw that in Ch. 4, we get

$$S_s \equiv -\left(\frac{\partial \widetilde{F}_s}{\partial T}\right)_{H_0} = S_n + \frac{VH_c}{4\pi}\frac{dH_c}{dT}. \qquad (4.11)$$

Since none of the terms on the right hand side depend on the field, the entropy of our specimen does not depend on the field (like in all other diamagnetics). This implies that n_s and therefore λ_L are *independent* of the field. The same conclusion can be drawn immediately from Eq. (5.61) and the MS definition. Following Pippard [128], the field independence of λ_L means that the long-range order of the superconducting phase extends over the entire volume of the specimen in the MS. Needless to say that this is exactly what follows from the MWM.

---

[54] No such a technique was available in 1950. Nowadays this can be done using Low-energy μSR (see, e.g., [116]).

[55] Pippard limited his analysis by the data at $T > 3$ K.

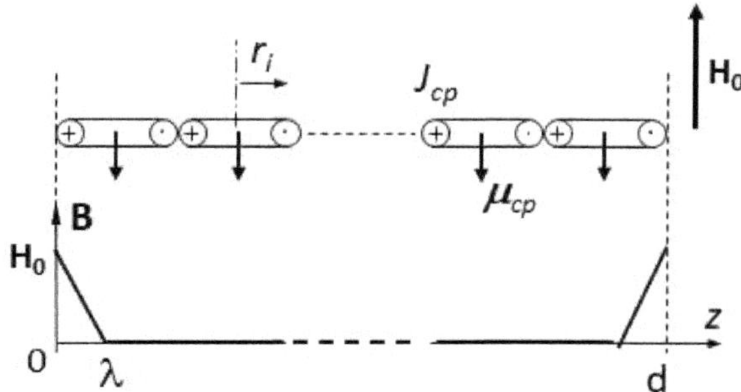

Figure 5.16 A cross section of the field-induced currents in Cooper pairs J_{cp} in a thin transverse (with respect to **H**) slice of a specimen in the MS as follows from the MWM. The specimen is a film with thickness d, like the film in Fig. 3.6. The slice contains currents induced in one Cooper pair of each micro-whirl of the specimen; r_i is their rms radius; and μ_{cp} is the magnetic moment induced in one pair. The graph schematically shows the distribution of induction **B** along the z axis perpendicular to the film surface; $\lambda(\neq \lambda_L)$ is the average width of the surface layer in which **B** decays from H_0 to zero. After Kozhevnikov [116], reprinted with permission from Springer Nature.

Thus, according to the thermodynamics argument of H. London, λ_L depends on the field, but the agreement of this statement with experiment is, at least, questionable. On the other hand, in accordance with experiment, the pure thermodynamic approach excludes such a dependence. However, it may look like that the latter approach ignores the presence of the penetration layer where $B \neq 0$, which is confirmed in countless experiments. Let us now turn to the MW model.

Fig. 5.16 schematically shows a cross section of a thin (≈ 6 fm thick) transverse slice of a specimen in the MS, as it follows from the MW model. The slice contains the induced currents in one Cooper pair J_{cp} of each micro-whirl of the specimen.

As one can see, in contrast to the London theory, there is no excluded volume. The currents induced in a unit volume just next to the boundary (at z near 0 and d) make the same contribution to the specimen magnetic moment as the induced currents in the unit volume in any other part of the specimen. Therefore, the magnetic moment per unit volume χ is the same everywhere inside the superconductor. Then, its magnetic moment is

$$M = \chi H V = -\frac{1}{4\pi} H V = -\frac{1}{4\pi} \frac{H_0}{(1 - \eta)} V,$$

where H is the field intensity, η is the demagnetizing factor, and V is the entire specimen volume as it was in the pure thermodynamic approach.

Hence, taking into account that r_i is the root mean square value, we arrive at the distribution of the averaged microscopic field, i.e., B, as schematically

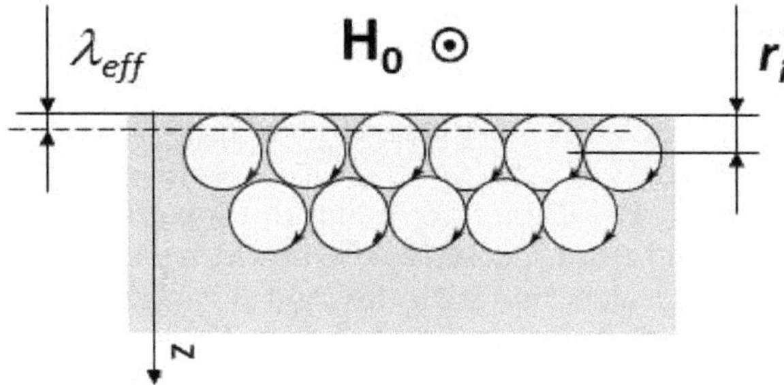

Figure 5.17   The field-induced currents (cross section of micro-whirls) near the specimen boundary in the MWM. The penetration field takes the gray areas at $z < r_i$. λ_{eff} is the effective penetration depth. The applied field H_0 is directed toward the reader.

shown in the graph at the bottom of Fig. 5.16. Thus, the MWM explains both the validity of the solely thermodynamic approach as well as the reality of the penetration layer.

Finishing our discussion of H. London's argument, we see that its erroneousness stems from the assumption of the screening current, which, in particular, leads to the appearance of a clearly non-physical excluded volume in this theory.

Finally, we have to address the question what is the penetration depth in MWM?

Cross sectional diagram of the field-induced currents near the specimen boundary in the MWM is schematically shown in Fig. 5.17. In the bulk, i.e. beyond the first row of the whirls, the induction is zero. However, as one can see from the figure, this is not the case at the edge, i.e. at the depth $\lesssim r_i$. Under the rough assumption that the induced currents are linear, the maximum penetration depth (λ) is equal to $r_i = 2\lambda_L$ and $\lambda_{eff} = \lambda_L/2$. This is, of course, an oversimplified estimate, however in any case, in the MWM λ is a finite value of the order of λ_L.

5.2.13   Type-I to type-II conversion, superconductivity of thin films and small particles

As known, the addition of an impurity to a flawless elementary type-I superconductor (e.g., alloying with another metal) converts it to a type-II superconductor and, correspondingly, leads to an increase of the critical field of the S/N transition denoting in this case as H_{c2}. As was for the first time shown by Shubnikov with coauthors [126], the higher the alloying percentage, the higher H_{c2}. Pippard reported that the alloying increases λ_L and reduces the mean free path [115]. The same kind of conversion takes place at decreasing the specimen dimensions, in particular the thickness of sufficiently thin films (see, e.g., [15]).

Pippard interpreted the type-I/type-II conversion as the result of a decrease in coherence length $\xi(= (1/\xi_0 + 1/l_m)^{-1})$, assumed to be a function of the mean free path l_m. Recall that according to Pippard, the wall energy parameter $\delta = \xi - \lambda_L$, and ξ is a spacial scale over which a change of F. London's ordering of the electron structure with the rigid $\widetilde{\mathbf{p}}$ can take place ("spreading tendency of the state of order" [207]). Correspondingly, the decreasing mean free path (which can also be expected in thin films) results in decreasing ξ and therefore can lead to the sign change of δ [115].

For the thin films an alternative interpretation of the increase of the critical field was proposed by H. London [138] and by Ginzburg and Landau near T_c [36].

In the MWM $\xi(\approx 2R_0)$ is the size of Cooper pair, a temperature-dependent constant of material determined by a polarization ability of its ionic lattice. At alloying, atoms of the other metal locally disturb the polarization of the lattice of the host material thus preventing the formation of pairs, which otherwise would exist at that locations. Apart from these places, the polarization stays unchanged. Correspondingly, the number density of Cooper pairs n_{cp} decreases, but R_0 is not altered. This leads to increase of λ_L, r_i and $\aleph$. The latter, being originally lesser than unity, at sufficient alloying (in a lead-tellurium alloy it can be as small as 0.8% of Tl [126]) becomes greater than one, meaning that the superconductor converts from type-I to type-II.

This picture is consistent with Pippard's observation of the increase of penetration depth at the alloying and with the fact that pure type-I superconductors can be converted to type-II ones, but not vice versa[56].

A similar in its essence process leads to an effective type-I/type-II conversion of the film samples at decreasing the film thickness d. When d becomes less than about $2R_0$ only Cooper pairs oriented so that $2R_0 \sin \phi < d$, where ϕ is the angle between $\boldsymbol{\mu}_0$ and the normal to the film, can survive. Respectively, the lesser d the lesser n_{cp} and the greater $\aleph$. This implies that at a definite thickness $\aleph$ becomes larger than unity and therefore the film behaves as a type II superconductor. Correspondingly, with a further decrease of the thickness the critical field becomes progressively greater than H_c.

The outlined interpretation is consistent with the well known fact that even very thin films of type-I materials remain superconducting and that these films always behave as type-II superconductors, as for the first time was revealed

---

[56]Type-I to type-II conversion can be also achieved by introducing structural defects. In such case the opposite conversion is possible by annealing.

in the classical experiment of Shalnikov [208][57] and confirmed in many other experiments afterward.

Possibly, the reader would like to ask why even ultrathin films (in Brickman's experiment on the London moment [205] the thinnest film was just $27\overset{\circ}{A}$ thick!) are superconducting and is there a thickness at which superconductivity ceases?

Coming from the MWM the answers can be as follows. The ultrathin films of a superconducting material superconduct (exhibit zero resistivity) when the field $\mathbf{H}$ and, therefore, the micro-whirls are parallel to the film. This was indeed so in the cited experiment of Shalnikov (the thinnest film was 5 nm thick) as well as in experiments on the London moment of Brickman (3 nm) [205] and Tate et al. (40 nm) [202]. At the same time one can expect a strong anisotropy in the superconducting properties of the ultrathin films and a small critical current, as it was indeed reported by Shalnikov.

One can imagine, that in the thinnest films the just mentioned angle ϕ is zero, i.e. the paired electrons circulate in the plane parallel to the film surface. However, the induced currents J_{cp}, which form the micro-whirls, lay in the planes transverse to $\mathbf{H}$ (see, e.g., Fig. 5.2). Quite obviously that in such a case r_i can be greater than the film thickness. But this should not bother us because we are aware that the whirls can travel through an insulating barrier of the Josephson junction. In particular, in the experiments on the London moment the film in Fig. 5.13 can be thinner than $2r_i$. At the same time all whirls are ordered and maintain the order in their translational (drift) motion forming the dissipation-free total current.

To address the second question, one can expect that even atomically thin films can be superconducting, provided that (a) they are metallic (the valence electrons are collectivized) and (b) the ionic lattice maintains the polarization necessary for the electron pairing. This suggests the possibility of the existence/synthesis of materials in which superconductivity is realized along certain atomic planes, as is indeed the case in multicomponent, called unconventional, superconductors (see, e.g., [209]).

Finally, the magnetic response of superconducting samples is unrelated to the whirl structure (see Sec. 5.2.2). This means that small particles of superconducting material (for example, particles of fine powder) can exhibit typical for superconductors magnetic properties. This fact is well known after the experiment of Shoenberg with colloidal samples [210].

---

[57]Shalnikov measured conductivity, critical field and critical current of tin and lead films with thickness from 5 to 200 nm. The films were deposited in high vacuum immediately in the cryostat (at 4.2 K) where all measurements were done. He found that even the thinnest films were superconducting; the critical field was much higher than that for the bulk metal, and increased with decreasing thickness; the critical current was much lower than in the bulk and decreased with the thickness; the Silsbee hypothesis failed.

5.2.14 Hall effect

As mentioned in Sec. 3.1.1, in one of the first experiments on superconductivity, Onnes and Hof revealed a non-existence of the Hall effect in Sn and Pb samples [82]. More specifically, they observed a sharp drop of the Hall voltage at crossing the critical temperature from above[58]. Based on this fact and also on results of his and Tune experiment of 1924 [79], Onnes suggested that superconducting electrons can be insensitive to the Lorenz force. Lorentz supported this interpretation ("...motion of the electrons is to a great extent insensible to the transverse forces exerted by the field" [83]). Hall's view was the opposite: "... electric currents have in one respect less freedom of motion in the supraconductive state than in the normal state" [65].

In the MWM the absence of the Hall effect naturally follows from the fact that the only action of the external magnetic field on the superconducting (paired) electrons (exerted through the field $\mathbf{H}$) is the change in frequency ω_i of their induced circular motion.

On the other hand, the absence of the Hall voltage generated by the normal electrons is explained by the fact that normal electrons do not take part in the total current in the S phase since the entire current (lesser of the critical current J_c) is transported by the ordered whirls.

We conclude this section with one more quote from Edwin Hall [65]: "The net Hall effect is the resultant of highly complicated conditions and actions in the normal state of metals. I am, however, of the opinion that supraconduction is a much simpler process than ordinary conduction". It is hard not to agree with these words.

5.2.15 Magnetic resonances and neutron scattering

Electron paramagnetic/spin resonance (EPR or ESR) and nuclear magnetic resonance (NMR) underlie the most powerful tools for studies of fine properties of matter in bulk: the ESR and NMR spectroscopy [46, 211]. Suffice to mention that in addition to physics, ESR and NMR are used in chemistry, biology and medicine (currently the main areas of application), as well as in geology, archaeology, food and other fields. Naturally, the ESR and NMR were also used in studies of superconductivity.

NMR. The nuclear magnetic resonance spectra were measured on samples of many elementary superconducting metals (see, e.g., [212, 213, 214, 216, 215]); later on experiments were performed on high-T_c and other multicomponent superconductors (see, e.g., [217]). In nearly all experiments the samples were colloidal particles, small platelets and fine grounded powder. Obtained experimental data on the Knight shift consistently demonstrate the failure of

---

[58]This was the first indication that superconductivity can differ from the perfect conductivity.

the standard theories, which was admitted, in particular, by the authors of the BCS theory [37, 38]. However, a quantitative interpretation of the experimental results was problematic due to a large width of the resonance lines and dependence of the results on the particle dimensions. The authors often attributed these difficulties to the smallness and scatter of particle sizes, and to the surface defects.

The choice of such kind of samples was due to the Londons' concept of the absence of a necessary for the resonance magnetic field in the bulk of massive superconductors. On the other hand, the dc field usually employed in NMR experiments greatly exceeds the critical field $H_{cr} = H_c$ in macroscopic type-I specimens, while small pieces are always type-II superconductors in which $H_{cr} = H_{c2} \gg H_{c1}$[59], which facilitates measurements.

However, inside superconductors the field intensity H is never zero. Therefore, since resonances are controlled by the H-field, both the NMR and ESR exist in massive specimens. Important that the latter can be made of the right (ellipsoidal) shape, which enables direct measurements of the H-filed. Apart from that, a great advantage of the massive specimens is that they avoid the distractions associated with the inevitable excess concentration of defects near the surface[60]. According to the MWM, at all temperatures the Knight shift in the S phase is caused by Cooper pairs *and* the normal electrons, herewith the contribution of the latter can be separated via the same measurements above H_{cr}.

ESR. Similar, but potentially even richer information about specifics of electron pairing can be obtained from ESR spectra. According to the MWM, superconductors may have two resonances caused by conduction electrons: the ESR due to precession of non-paired electrons (the resonance frequency $\nu_{esr} = eH/mc$, $g = 2$) and a resonance due to the precession of Cooper pairs (let us call it as electron diamagnetic resonance, EDR) at $\nu_{edr} = eH/2mc$, $g = 1$.

As mentioned in Sec. 5.2.8, a possibility of the EDR can be expected based on the existence of the direct and reversed Kikoin-Goobar effects. Available in literature an experimental report on ESR in a pure superconductor (Nb) [219] is related to the mixed state; the data demonstrate existence of ESR, but due to the specimen magnetic inhomogeneity, it is impossible to identify either the resonance occurs in the S or/and N phase(s); the frequency used in this research (8.4 GHz) was too high for testing the EDR.

SANS. The small angle neutron scattering has never been reported for superconductors in the Meissner state. However, such experiments (preferably with polarized neutrons) potentially can allow one to conduct direct measurements of the parameters of the whirl structure.

---

[59]For example, in the massive tin specimens $H_{cr} = H_{c1} = H_c \lesssim 300$ Oe, whereas tin particles used by Andreas and Knight $H_{cr} = H_{c2} = 25000$ Oe [216].

[60]Pauli once said "God made the bulk; surfaces were invented by the devil" [218].

5.3   PROBLEMS

5.1. Derive the formula for j_{cr} (5.31).
Hint. Calculate v_{ic}^2 from (4.6) and use it to calculate $j_{cr}^2 = [n_{cp}(2e)v_{ic}]^2$.

5.2. Show that the flux passing through a "hole" appearing when super-conductivity ceases in a single micro-whirl is equal to the flux quantum Φ_0. Consider a cylindrical specimen with base area A and length L.
Solution. Referring to Sec. 5.2.2. for denotations, we write

$$N_{cp} = n_{cp}V = (n_\perp A)(n_\parallel L),$$

where $n_\perp$ is the number of micro-whirls per unit area.
According to the flux quantization, $\Phi_0 n_\perp A = HA$, so

$$\Phi_0 = \frac{Hn_\parallel}{n_{cp}} = \frac{H}{n_{cp}} \cdot \frac{mc^2}{2e^2} = H\lambda_L^2 4\pi = H(\pi r_i^2).$$

Hence, when superconductivity ceases in one micro-whirl, the flux passing through the appeared hole is

$$B\pi r_i^2 = H\pi r_i^2 = \Phi_0.$$

5.3. What is the energy per unit length (line tension e_l) of a flux line passing a specimen of type-II superconductor? What is e_l at $H_0 = H_{c1}$? Consider the specimen of cylindrical geometry.
Solution.

$$e_l = \frac{E_l}{L} = \frac{1}{L}\left[\frac{H^2}{8\pi}\pi r_i^2 L + E_k\right] = \frac{1}{L}\left[\frac{H^2}{8\pi}\pi r_i^2 L + \frac{H^2}{8\pi n_{cp}}n_\parallel L\right] =$$
$$\frac{1}{L}\left[\frac{H^2}{8\pi}\pi r_i^2 L + \frac{H^2}{8\pi n_{cp}}4\pi n_{cp}\lambda_L^2 L\right] = \frac{H^2}{4\pi}\pi r_i^2 = \frac{H\Phi_0}{4\pi},$$

where E_k is kinetic energy of the induced motion of the bound electrons at the flux line boundary, as shown in Fig.5.12. At $H_0 = H_{c1}$ the line tension is

$$e_i(H_c1) = \frac{H_{c1}\Phi_0}{4\pi}.$$

This formula is the same as that obtained in the standard theory with the Abrikosov vortices, in which the first vortex (the one appearing at H_{c1}) does not have neighbors [167].

5.4. A long niobium cylinder is cooled to 2 K and rotates about its longitudinal axis with angular velocity $\omega = 100$ rps. The cylinder is in free space; its radius is $R = 0.5$ cm and the length is $L = 10$ cm; the lower critical field of niobium at 2 K $H_{c1} = 1100$ Oe. Find: (a) the intensity H and (b) the induction B of the magnetic field inside it, (c) H and B outside the cylinder; (d) the cylinder magnetic moment.

Solution.

(a) $\mathbf{H} = -(2mc/e)\boldsymbol{\omega} = 7.15 \cdot 10^{-5}$ Oe; (b) since $H < H_{c1}$, $\mathbf{B} = 0$; (c) due to the continuity of H_t, everywhere outside the cylinder $H_{ext} = 7.15 \cdot 10^{-5}$ Oe; and $B_{ext} = H_{ext} = 7.15 \cdot 10^{-5}$ G; (d) magnetic moment per unit length $\mathbf{M}/L = -(R^2/4)\mathbf{H} = -4.5 \cdot 10^{-6}$ EMU/cm or erg·(G cm)$^{-1}$.

5.5. Answer the same questions as in the previous problem for niobium ball of $R = 0.5$ cm.

Solution.

(a) and (b), the same answers. (c) Due to the continuity of B_n, the outer field near the ball bends around it as in Fig. 3.10. The outside field near ball $B_{ext} = H_{ext} = H_{\parallel} = H\sin\theta$ (see problem 1.4); away from the ball $H_{ext} = H(1 - \eta) = 2H/3 = 4.8 \cdot 10^{-5}$ Oe, and $B_{ext} = 4.8 \cdot 10^{-5}$ G. (d) The magnetic moment $\mathbf{M} = -(V/4\pi)\mathbf{H} = -(R^3/2)\mathbf{H} = -4.5 \cdot 10^{-6}$ EMU.

5.6. Answer the same questions as in the previous two problems for a niobium disc with a diameter $D = 50$ cm and thickness $d = 0.5$ cm.

Solution.

Since $d \ll D$, the demagnetizing factor of this disc η can be taken equal to unity. (a) and (b), the same answers. (c) Due to the continuity of B_n, the outer field (both B and H) is zero. The disk can be considered as a strongly flattened (compressed) ball with the field lines moved far apart. (d) $\mathbf{M} = -(V/4\pi)\mathbf{H} = 0.07$ EMU.

Note the lack of analogy between this rotation-magnetized disk and the same stationary disk with an applied field. In the latter case such a disc (with $\eta = 1$) can never be in the MS, because the lines of the applied field must go though the disc and therefore it can be found only in the MXS (or in the IS, if it is made of a type-I superconductor).

Bibliography

[1] J. C. Maxwell, *A Treatise on Electricity and Magnetism*, v.II, 2nd ed. (Clarendon Press, Oxford, 1873).

[2] E. A. Guggenheim, Proc. Roy. Soc. A (London) **155**, 49 (1936).

[3] E. A. Guggenheim, Proc. Roy. Soc. A (London) **155**, 70 (1936).

[4] H. Kamerlingh Onnes, KNAW Proceedings: **13**, 1274 (1911); **14**, 113 (1911); **14**, 818 (1912).

[5] W. Meissner, Z. ges. Kälteindustr. **34**, 197 (1927).

[6] C. J. Gorter and H. B. G. Casimir, Phys. Z. **35**, 963 (1934).

[7] W. Meissner and R. Ochsenfeld, Naturwissenschaften **21**, 787 (1933).

[8] G. N. Rjabinin and L. W. Shubnikow, Nature **134**, 286 (1934).

[9] J. G. Bednorz and K. A. Müller, Z. Physik B **64**, 189 (1986).

[10] L. P. Gorkov and V. Z. Kresin, Rev. Mod. Phys. **90**, 011001 (2018).

[11] L. N. Cooper, Phys. Rev. **104**, 1189 (1956).

[12] B. S. Deaver, Jr., and W. M. Fairbank, Phys. Rev. Letters **7**, 43 (1961).

[13] R. Doll and M. Näbauer, Phys. Rev. Lett. **7**, 51 (1961).

[14] A. B. Pippard, Proc. Roy. Soc. A, **203**, 98 (1950).

[15] D. Shoenberg, *Superconductivity*, 2nd. ed., (Cambridge, University Press, 1962).

[16] V. Kozhevnikov, Thermodynamics of Magnetizing Materials and Superconductors (CRC Press, Boca Raton, 2019).

[17] J. D.Jackson, The nature of intrinsic magnetic dipole moments (CERN-77-17). European Organization for Nuclear Research (CERN), 1977.

[18] I. E. Tamm, *Fundamentals of the Theory of Electricity* 9th ed. (Mir, Moscow, 1979).

[19] D. J. Griffiths, *Introduction to Electrodynamics*, 5th ed. (Cambridge University Press, Cambridge, 2023).

[20] L. D. Landau, E.M. Lifshitz and L. P. Pitaevskii, *Electrodynamics of Continuous Media*, 2nd ed. (Elsevier, 1984).

[21] H. A. Lorentz, *The theory of Electrons* (The Columbia University Press, New York, 1909).

[22] E. M. Purcell, *Electricity and Magnetism*, 2nd. Ed. (Mc-Graw Hill, Boston, 1985).

[23] J. H. Van Vleck, *The theory of electric and magnetic susceptibilities* (Clarendon Press, Oxford, 1932).

[24] V. K. Pecharsky, K. A. Gschneidner Jr., in Encyclopedia of Condensed Matter Physics, 2nd ed., T. Chakraborty Ed., v.2, p.85 (Elsevier, Amsterdam, 2023).

[25] Par M. Poisson, Memoire sur La Theorie Du Magnetisme, Lu a l'Academie Royale des Sciences, 2 Fevrier, 1824.

[26] I. S. Grigoriev, E. Z. Meilikhov and A. A. Radzig, Eds., *Handbook of Physical Quantities* (CRC Press, Boca Raton, 1997).

[27] V. S. Egorov, G. Solt,, C. Baines, D. Herlach and U. Zimmermann, Phys. Rev. B **64**, 024524 (2001).

[28] V. J. Katz, Mathematics Magazine **52**, 146 (1979).

[29] J. A. Osborn, Phys. Rev. **67**, 351 (1945).

[30] V. Kozhevnikov and C. Van Haesendonck, Phys. Rev. B **90**, 104519 (2014).

[31] L. D. Landau and E. M. Lipschitz, The Classical Theory of Fields, 4th ed. (Elsevier, 1975).

[32] L. D. Landau and E. M. Lifshitz, *Quantum Mechanics*, 3d Ed. (Elseivier Science Ltd., Amsterdam, 1977).

[33] W. Ehrenberg and R. E. Siday, Proc. Roy. Soc. B **62**, 8 (1949).

[34] Y. Aharonov and D. Bohm, Phys. Rev. **115** 485 (1959).

[35] F. London, *Superfluids* v. I (Dover, N.Y., 1960).

[36] V. L. Ginzburg and L. D. Landau, Zh.E.T.F. **20**, 1064 (1950).

[37] J. Bardeen, L. N. Cooper and J. R. Schrieffer, Phys. Rev. **108**, 1175 (1957).

[38] J. R. Schrieffer, *Theory of Superconductivity*, (Taylor & Francis Group, Boca Raton, 1999).

[39] F. Bloch, *Molekulartheorie des Magnetismus* (Akademische Verlagsgesellschaft, Leiptzig, 1934).

[40] C. Kittel, Introduction to Solid State Physics, 8th ed. (John Wiley $ Sons, Inc., NJ, 2005).

[41] N. W. Ashcroft and N. D. Mermin, Solid State Physics (Brooks/Cole Cengage Learning. Belmont, USA, 1976).

[42] *Perspectives in Quantum Hall Effects*, Eds. S. Das Sarma and A. Pinczuk, (Waley, N.Y., 1997).

[43] L. D. Landau and E. M. Lifshitz, *Statistical Physics*, Part I, 3d ed. (Elsevier, Amsterdam, 1980).

[44] L. D. Landau, E.M. Lifshitz *Mechanics*, 3d ed. (Elsevier, 1976).

[45] I. K. Kikoin and S. V. Goobar, C. R. Acad. Sci. USSR **19**, 249 (1938); J. Phys. USSR **3**, 333 (1940).

[46] A. Abragam and B. Bleaney, *Electron Paramagnetic Resonance of Transition Ions* (Calendon Press, Oxford, 1970).

[47] S. Blundel, Magnetism in Condensed Matter (Oxford University Press, Oxford, 2001).

[48] J. Larmor, The London, Edinburgh, Dublin Phil. Mag. and J. Science 44, 503 (1897).

[49] R. Feynman, R. Leighton, M. Sands, *The Feynman Lectures on Physics*, v. II (Basic Books, N.Y., 1964).

[50] E. V. Shpol'skii, Atomic physics, v.I, (Iliffe Books, London, 1969).

[51] M. Faraday, *Experimental Researches in Electricity* v. III, Phil. Trans. Roy. Soc. (London) **136**, 41 (1846).

[52] *Wilhelm Weber's main works on electrodynamics translated into English*, A. K. T. Asis. Ed. (C. Roy Keys Inc, Montreal, 2021).

[53] S. J. Barnett, Phys. Rev. 6, 239 (1915).

[54] A. Einstein, W. J. de Haas, KNAW Proceedings, **18**, 696 (1915).

[55] S. J. Barnett. Rev. Mod. Phys. **7**, 129 (1935).

[56] A. K. T. Assis and J. P. M. C. Chaib *Ampère's Electrodynamics* (C. Roy Keys Inc., Montreal 2015).

[57] O. V. Auwers, Naturwissenschaften **23**, 202 (1935).

[58] P. L. Kapitza, Experiment, Theory, Practice (D. Reidel Pub. Com., Dordreht, Holland, 1980).

[59] J. Perry, *Spinning Tops. The "Operatives' Lecture" of the British Association Meeting at Leeds, 6th September 1890*; available online at https://www.gutenberg.org/cache/epub/34268/pg34268-images.html.

[60] O. W. Richardson, Phys. Rev. (Series I) **26**, 248 (1908).

[61] F. Coeterier and P. Scherrer, Helv. Phys. Acta **5**, 217 (1932).

[62] I. K. Kikoin, Zh. Tekh. Fiz. (Technical Physics) **166**, 129 (1946); reprinted in *I. K. Kikoin - Physics and Fate*, Ed. S. S. Yakimov, p. 145 (Nauka, Moscow, 2008); online version: http://elib.biblioatom.ru/text/kikoin-fizika-i-sudba_ 2008/go,112/.

[63] R. H. Pry, A. L. Lathrop, and W. V. Houston, Phys. Rev. **86**, 905 (1952).

[64] A. F. Hildebrandt, Phys. Rev. Letters **12**, 190 (1964).

[65] E. H. Hall, Proc. Nat. Acad. Sci. Wash. **19**, 619 (1933).

[66] P. Langevin, Comptes rendus de l'Academie des Sciences **139**, 1204 (1905).

[67] P. Langevin Annales de Chimie etcitekey de Physique **5**, 70 (1905).

[68] P. Langevin, Journal de Physique **4**, 678 (1905).

[69] L Navarro, J Olivella, Archives internationales d'histoire des sciences **47**, 316, (1997).

[70] W. Pauli jr., ZS. Phys. **2**, 201 (1920).

[71] P. W. Selwood, *Magnetochemistry*, 2nd Ed. (Interscience Publishers Inc., N.Y., 1956).

[72] P. Ehrenfest, Physica **5**, 388 (1925).

[73] C. V. Raman, Nature **124**, 412 (1929).

[74] E. K. Zavoiski, J. Phys. (USSR) **9**, 245 (1945).

[75] W. E. Henry, Phys. Rev. **88**, 559 (1952).

[76] D. van Delft and P. Kes, Physics Today, Sept. 2010, p. 38.

[77] D. van Delft, *Freezing Physics. Kamerlingh Onnes and the Quest for Cold*, (KNAW, Amsterdam, 2007).

[78] B. T. Matthias, T. H. Geballe and V. B. Compton, Rev. Mod. Phys. **35**, 1 (1963).

[79] H. Kamerlingh Onnes, Phys. Lab. Univ. Leiden, Suppl. No 50a (1924).

[80] T. Sauer, Archive for History of Exact Sciences **61**, 159, 2006.

[81] H. G. Smith and J. O. Wilhelm, Rev. Mod. Phys. **7**, 237 (1935).

[82] H. K. Onnes, K. Hof, Proc. Roy. Soc. Amsterdam **17**, 520 (1914).

[83] H. A. Lorentz, Phys. Lab. Univ. Leiden, Suppl. No 50b (1924).

[84] A. Einstein, in *Het natuurkundig laboratorium der Rijksuniversiteit te Leiden in de jaren 1904–1922. Gedenkboek aangeboden aan H. Kamerlingh Onnes, Directeur van het laboratorium, bij gelegenheid van zijn veertigjarig professoraat op 11 November 1922.*, p. 429 (Leiden, Eduardo Ijdo, 1922); English translation: arXiv:physics/0510251.

[85] M. Tinkham, Introduction to Superconductivity, 2nd ed. (McGraw-Hill, NY, 1996).

[86] D. Dew-Hughes, Low Temp. Phys. **27**, 713 (2001).

[87] J. File and R. G. Mills, Phys. Rev. Letters **10**, 93 (1963).

[88] J. File, J. Appl. Phys. **39**, 2335 (1968).

[89] J. Mehra, *The Sovay Conferences on Physics* (D. Redel Publishing Co, Dordrecht-Holand, 1975).

[90] G. Borelius W. H. Keesom, C. H. Johansson and J. O. Linde, KNAW Proceedings **35**, 10 (1932).

[91] N. Cusack and P. Kendall, Proc. Phys. Soc. **72** 898 (1958).

[92] J. W. Christian, J.-p Jan, W. B. Pearson and I. M. Templeton, Proc. R. Soc. Lond. A**245**, 213 (1958).

[93] H. B. Gallen, Thermodynamics and an Introduction to Thermostatics, 2nd ed., (John Wiley & Sons, N.Y., 1985).

[94] V. L. Ginzburg, Zh.E.T.F. **14** 177 (1944); J. Phys. USSR **8** 148 (1944).

[95] W. J. de Haas and H. Bremmer, KNAW Proc. **34**, 325-338 (1931).

[96] W. H. Keesom, J. N. van den Ende, KNAW Proc. **35**, 143 (1932).

[97] W. H. Keesom and J. A. Kok, KNAW Proc. **35**, 743 (1932).

[98] W. H. Keesom and J. A. Kok, Commun. Phys. Lab. Univ. Leiden, No 230c (1932).

[99] P. F. Dahl, *Superconductivity: its historical roots and development from mercury to the ceramic oxides* (American Institute of Physics, N.Y., 1992).

[100] W. Meissner and F. Heidenreich, Phys. Z. **37**, 451 (1936).

[101] W. Meissner and F. Heidenreich, Phys. Z. **37**, 455 (1936).

[102] W. Meissner, R. Ochsenfeld and F. Heidenreich, Zeitsch. Gesamt. Kalte-Industr. **41**, 125 (1934).

[103] K. Mendelssohn and J. D. Babbitt, Nature **133**, 459 (1934).

[104] H. Grayson Smith and J. O. Wilhelm, Proc. Roy. Soc. A **157**, 132 (1936).

[105] F. G. A. Tarr and J. O. Wilhelm, Trans. Roy. Soc. Canada **28**, 61 (1934); Can. J. Research **12**, 265 (1935).

[106] K. Mendelssohn and J. D. Babbitt, Proc. R. Soc. A **151**, 316 (1935).

[107] V. Kozhevnikov, A. Suter, T. Prokscha, C. Van Haesendonck, J. Supercond. Nov. Magnetism **33**, 3361 (2020).

[108] D. Schoenberg, Proc. Roy. Soc. A **155**, 712 (1936).

[109] L. W. Schubnikow ans W. I. Chotkewitsch Phys. Z. Sowiet. **10**, 231 (1936).

[110] A. I. Shal'nikov, Zh. Eksp. Teor. Fiz. **33**, 1071 (1957) [English translation: JETP **6**, 827 (1958)].

[111] A. B. Pippard, J. G. Shepherd and D. A. Tindall, Proc. Roy. Soc. A. **324**, 17 (1971).

[112] W. W. Parker, *Self Induction Formulas*, Thesis, University of Illinois, 1908.

[113] J. A. Kok, Physica **1**, 1103 (1934).

[114] F. H. London, Proc. Roy. Soc. A **149**, 71 (1935).

[115] A. B. Pippard, Proc. Roy. Soc. A **216**, 547 (1953).

[116] V. Kozhevnikov, J. Supercond. Nov. Magn. **34**, 1979 (2021).

[117] R. Becker, G. Heller and F. Sauter, Z. Phys. **85**, 772 (1933).

[118] A. Suter, E. Morenzoni, N. Garifianov, R. Khasanov, E. Kirk, H. Luetkens, T. Prokscha, and M. Horisberger, Phys. Rev. B **72**, 024506 (2005).

[119] V. Kozhevnikov, A. Suter, H. Fritzsche, V. Gladilin, A. Volodin, T. Moorkens, M. Trekels, J. Cuppens, B. M. Wojek, T. Prokscha, E. Morenzoni, G. J. Nieuwenhuys, M. J. Van Bael, K. Temst, C. Van Haesendonck, J. O. Indekeu, Phys. Rev. B **87**, 104508 (2013).

[120] L. Onsager, Phys. Rev. Letters **7**, 50 (1961).

[121] J. D. Jackson, *Classical Electrodynamics*, 3d ed. (John Wiley & Sons, Inc., Hoboken NJ, 1999).

[122] V. Kozhevnikov, in *Encyclopedia of Condensed Matter Physics*, 2nd. ed., v. 2, p. 644, p. (Elsevier, Amsterdam, 2023).

[123] B. D. Josephson. Phys. Rev. Letters **1**, 251 (1962).

[124] L. P. Gorkov, in *100 Years of Superconductivity*, p. 72, Eds. H. Rogalla and P. H. Kes (CRC Press, Boca Raton, 2012).

[125] H. Koppe and J. Willebrand, J. Low Temp. Phys. **2**, 499 (1970).

[126] L. V. Shubnikov, V. I. Khotkevitch, Yu. D. Shepelev, Yu. N. Ryabinin, Zh.E.T.F. Zh. Eksper. Teor. Fiz. **7**, 221 (1937).

[127] A. A. Abrikosov, Sov. Phys. JETP **5**, 1174 (1957).

[128] A. B. Pippard, Proc. Roy. Soc. A, **203**, 210 (1950).

[129] A. B. Pippard, in Proc. NATO Advanced Study Institute on Small-Scale Applications of Superconductivity, Gardone, 1976, B. B. Schwartz, S. Foner, eds., Plenum, New York (1977); reprinted in *100 Years of Superconductivity*, H. Rogalla and P. Kess eds., p. 29 (CRC Press, Boca Baton, 2012).

[130] V. Kozhevnikcv, C. Van Haesendonck, T. Prokscha, A.-M. Valente-Feliciano, 15th Int. Conf. on Muon Spin Rotation, Relaxation and Resonance, Abstract book, p. 182, Parma 2022.

[131] L. N. Cooper, in *BCS: 50 Years*, p. 3 (World Scientific Pub. Com., Singapore 2010).

[132] J. Bardeen, Phys. Rev. **97**, 1724 (1955).

[133] E. Maxwell, Phys. Rev. **78**, 477 (1950).

[134] C. A. Reynolds, B. Serin, W. H. Wright, and L. B. Nesbitt, Phys. Rev. **78**, 487 (1950).

[135] B. T. Matthias, Science **144**, 378 (1964).

[136] V. Z. Kresin and S. A. Wolf, *Fundamentals of Superconductivity*, (Premium Press, N. Y., 1990).

[137] L. D. Landau, Zh. Eksp. Teor. Fiz. **13**, 377 (1943) [English translation: J. Phys. USSR **7**, 99 (1943)].

[138] H. London, Proc. Roy. Soc. A (London) **152**, 650 (1935).

[139] A. B. Pippard, Mathem. Proc. Cambridge Phil. Soc. **47**, 617 (1951).

[140] I. T. Faber, Proc. Roy. Soc. A **248**, 460 (1958).

[141] H. F. Hess, R. B. Robinson, R. C. Dynes, J. M. Valles, Jr., and J. V. Waszczak, Phys. Rev. Letters **62**, 215 (1989).

[142] J. Beare, M. Nugent, M.N. Wilson, Y. Cai, T.J.S. Munsie, A. Amon, A. Leithe-Jasper, Z. Gong, S.L. Guo, Z. Guguchia, Y. Grin, Y.J. Uemura, E. Svanidze and G.M. Luke, Phys. Rev. B **99**, 134510 (2019).

[143] D. K. Finnemore and D. E. Mapother, Phys. Rev. **140**, A507 (1965).

[144] V. Kozhevnikov, R. J. Wijngaarden, J. de Wit, and C. Van Haesendonck, Phys. Rev. B **89**, 100503 (2014).

[145] A. J. Rutgers, Physica **1**, 1055 (1934).

[146] P. Ehrenfest, Comm. Leiden Suppl. No 75b (1933); English translation in: T. Sauer, The European Physical Journal Special Topics **226**, 539 (2017).

[147] C. J. Gorter and H. B. G. Casimir, Physica **1**, 305 (1934).

[148] L. Shubnikov and I Nikhutin, Nature **139**, 589 (1937).

[149] L. J. Reinders, *The Life, Science and Times of Lev Vasilevich Shubnikov*, Springer International Publishing AG 2018.

[150] R. P. Huebener, *Magnetic flux structures in Superconductors*, 2nd ed. (Springler-Verlag, Berlin, 2010).

[151] R. Peierls, Proc. Roy. Soc. A **155**, 613 (1936).

[152] F. London, Physica **3**, 450 (1936).

[153] M. Desirant and D. Shoenberg, Proc. Roy. Soc. A **194**, 63 (1948).

[154] L. D. Landau, Zh. Eksp. Teor. Fiz. **7**, 371 (1937) [English translation: Phys. Zs. Sowjet. **11**, 129 (1937)].

[155] L. D. Landau, Nature **141**, 688 (1938).

[156] A. G. Meshkovsky and A. I. Shalnikov, Zh.E.T.F. **17**, 851 (1947).

[157] Y. V. Sharvin, Zh.E.T.F. **33**, 1341 (1957) [Sov. Phys. JETF **6**, 1031 (1958)].

[158] A. A. Abrikosov, *Fundamentals of the Theory of Metals* (Elsevier, Amsterdam, 1988).

[159] V. Kozhevnikov, A. Suter, T. Prokscha, C. Van Haesendonck, J. Supercond. Nov. Magnetism **33**, 3361 (2020).

[160] V. Kozhevnikov, A.-M. Feliciano, P. J. Curran, G. Richter, A. Volodin, A. Suter, S. J. Bending, C.Van Haesendonck, J Supercond. Nov. Magn. **31**, 3433 (2018)

[161] U. Essmann and H. Träuble, Phys. Letters **24A**, 526 (1967).

[162] D. K. Finnemore, T. F. Stronberg and C. A. Swenson, Phys. Rev. **149**, 231 (1966).

[163] A.-M. Valente-Feliciano, Development of SRF monolayer/multilayer thin film materials to increase the performance of SRF accelerating structures beyond bulk Nb, PhD dissertation, Université Paris Sud - Paris XI, 2014.

[164] P. G. de Gennes, *Superconductivity of Metals and Alloys* (Perseus Book Publishing, L.L.C., 1966).

[165] C. P. Bean and J. D. Livingston, Phys. Rev. Letters **12**, 14 (1964).

[166] D. Saint-James and P. G. De Gennes, Phys. Letters **7**, 306 (1963).

[167] E. M. Lifshitz and L. P. Pitaevskii *Statistical Physics* v.2, (M., Nauka, 1973).

[168] V. Kozhevnikov, A.-M. Valente-Feliciano, P. J. Curran, A. Suter, A. H. Liu, G. Richter, E. Morenzoni, S. J. Bending, and C. Van Haesendonck, Phys. Rev. B **95**, 174509 (2017).

[169] V. Kozhevnikov, in *Encyclopedia of Condensed Matter Physics*, 2nd. ed., v.2, p. 644 (Elsevier, Amsterdam, 2023).

[170] L. N. Cooper, Nobel lecture, 1972.

[171] L. B. Okun, Physics-Uspekhi **178**, 653 (2008); arXiv:0809.2379 (2008).

[172] R. P. Feynman, Progr. Low Temp. Phys. **I**, 17 (1955).

[173] E. J. Yarmchuk and R. E. Packard, J. Low Temp. Phys. **46**, 479 (1982).

[174] H. K. Onnes, Communications of the Physical Laboratory of Leiden University, No 133a and 133b (1913).

[175] H. K. Onnes, Communications of the Physical Laboratory of Leiden University, No 133d (1913).

[176] H. K. Onnes, Communications of the Physical Laboratory of Leiden University, No 139f (1914).

[177] F. B. Silsbee, Journal of the Washington Academy of Sciences **6**, 79 (1916).

[178] W. Tuyn and H. K. Onnes, J. Franklin Inst. **201**, 379 (1926).

[179] L. W. Shubnikov and N. E. Alexejevski, Nature **138**, 804 (1936).

[180] N. E. Alexejevski, Zh.E.T.F. **8**, 342 (1938).

[181] R. B. Scott, J. Research Natl. Bur. Standards **41**, 581 (1948).

[182] W. H. Keesom, Communication No. 234f from the Kamerlingh Onnes Laboratory at Leiden, 1936.

[183] E. Zeldov, in *100 Years of Superconductivity*, Eds. H. Rogulla and P. H. Kes, p. 222 (CRC Press, Roca Raton, 2012).

[184] A. Polyanskii, private communication (to be published).

[185] A. Barone and G. Paterno, *Physics and Applications of the Josephson Effect* (John Wiley & Sons, N.Y., 1982).

[186] A. Shoji, in Encyclopedia of Materials: Science and Technology, p. 4347 (Elsevier, Amsterdam, 2001).

[187] P. Durandetto, D. Serazio, A. Sosso, Acta IMEKO, vol. 12, no. 3, art. 19 (2023).

[188] K. K. Likharev, *Dynamics of Josephson Junctions and Circuits* (CRC Press, Boca Baton, 1986).

[189] *Fundamentals and Frontiers of the Josephson Effect*, Ed. F. Tafuri (Springer Nature, Switzerland, 2019).

[190] D. Roditchev, C. Brun, L. Serrier-Garcia, J. Cuevas, V. Bessa, M. Milošević, F.Debontridder, V. Stolyarov and T. Cren, Nature Physics **11**, 332 (2015).

[191] Matthias Micha Wildermuth, Quantum *Tunneling of Josephson Vortices in High-Impedance Long Junctions* (KIT Scientific Publishing, Karlsruhe, 2023).

[192] B. D. Josephson, Rev. Mod. Phys. **16**, 216 (1964).

[193] F. Bloch, Phys. Rev. Letters **21**, 1241 (1968).

[194] S. Shapiro, Phys. Rev. Letters **11**, 80 (1963).

[195] W. Thomson, Trans. Roy. Soc. Edinburgh **XXI**, 123 (1853).

[196] M. W. Zemansky and R. H. Dittman, *Heat and Thermodynamics*, 7th Ed.,(McGraw-Hill Inc, NY, 1997).

[197] K. Behnia, Fundamentals of Thermoelectricity, (Oxford University Press, Oxford, 2015).

[198] R. Becker, G. Heller and F. Sauter, Z. Phys. **85**, 772 (1933).

[199] E. A. Lynton, *Superconductivity*, 2nd ed. (Methuen, London, 1964).

[200] H. Ess'en, Eur. J. Phys. **26**, 279 (2005).

[201] J. E. Hirsch Phys. Scr. **89**, 015806 (2014).

[202] J. Tate, B. Cabrera, S. B. Felch, and J. T. Anderson, Phys. Rev. Letters **62**, 845 (1989).

[203] A. A. Verheijen, J. M. van Ruitenbeek, R. de Bruyn Ouboter and L. J. de Jongh, Nature **345**, 418 (1990).

[204] S. Buchman, C. W. F. Everitt , B. Parkinson, J. P. Turneaure and G. M. Keiser, Physica **B** 280 497 (2000).

[205] N. F. Brickman, Phys. Rev. **184**, 460 (1969).

[206] H. London, Physical Society Report on a Low Temperature Conference, London, p. 51 (1947), cited from [35].

[207] A. B. Pippard, IEEE Trans. Magnetics **23**, 371 (1987).

[208] A. Shalnikov, Nature **142**, 74 (1938).

[209] G. R. Stewart, Advances in Physics **66**, 75 (2017).

[210] D. Shoenberg, Proc. Roy. Soc. A **175**, 49 (1940).

[211] A. Abragam and M. Goldman, *Nuclear Magnetism: Order and Disorder* (Oxford University Press, Oxford, 1982).

[212] W. D. Knight, G. M. Androes, and R. H. Hammond, Phys. Rev. **104**, 852 (1956).

[213] F. Reif, Phys. Rev. **106**, 208 (1957).

[214] L. C. Hebel and C. P. Slichter, Phys. Rev. **107**, 901 (1957).

[215] W. A. Hines and W. D. Knight, Phys. Rev. Lett. **18**, 341 (1967).

[216] G. M. Androes and W. D. Knight, Phys. Rev. **121**, 779 (1961).

[217] K. Matano, S. Maeda, H. Sawaoka, Y. Muro, T. Takabatake, B. Joshi, S. Ramakrishnan, K. Kawashima, J. Akimitsu, and G. Zheng, J. Phys. Soc. Jpn. **82**, 084711 (2013).

[218] M. Schroeder, *Fractals, Chaos, Power Laws: Minutes from an Infinite Paradise* (M. H. Freeman & Co, 1991).

[219] Y. Yafet, D. C. Vier and S. Schultz, J. Appl. Phys. **55**, 2022 (1984).

Index